LABORATORY EXPERIMENTS

Seventh Edition

BASIC CHEMISTRY

LABORATORY EXPERIMENTS

Charles H. Corwin
Department of Chemistry
American River College

---- *Seventh Edition* ----

BASIC CHEMISTRY

G. WILLIAM DAUB
Department of Chemistry
Harvey Mudd College

WILLIAM S. SEESE
Department of Chemistry
Casper College

Prentice
Hall

PRENTICE HALL, UPPER SADDLE RIVER, NJ 07458

Production Editor: *Carole Suraci*
Production Supervisor: *Joan Eurell*
Acquisitions Editor: *Ben Roberts*
Supplement Acquisitions Editor: *Mary Hornby*
Production Coordinator: *Ben Smith*

©1996 by Prentice-Hall, Inc.

Upper Saddle River, New Jersey 07458

Printed in the United States of America

15 14

ISBN: 0-13-378506-8

Prentice-Hall International (UK)Limited, *London*
Prentice-Hall of Australia Pty. Limited, *Sydney*
Prentice-Hall Canada Inc., *Toronto*
Prentice-Hall Hispanoamericana, S.A., *Mexico*
Prentice-Hall of India Private Limited, *New Delhi*
Prentice-Hall of Japan, Inc., *Tokyo*
Pearson Education Asia Pte. Ltd., *Singapore*
Editora Prentice-Hall do Brasil, Ltda., *Rio de Janerio*

Contents

*Assigned Student Unknowns.

APPENDICES

Preface

TO THE INSTRUCTOR:

Your critical responses to *Laboratory Experiments: Basic Chemistry, Seese/Daub, 6/e* provided many valuable comments for this revised seventh edition. The previous edition's format has been retained with the following components: Objectives; Discussion including problem examples, diagrams and illustrations; list of Equipment and Chemicals; a stepwise Procedure; a Prelaboratory Assignment with answers in the Appendix; Data Tables for recording observations and measurements; and a Postlaboratory Assignment designed to synthesize the laboratory experience.

Several basic laboratory techniques, such as using a burner, balance, or pipet, are found in the *Appendices*. Also in the appendices is a glossary of key terms, answers to the prelaboratory assignments, and applicable reference tables. *Laboratory Experiments, Basic Chemistry, 7/e* includes the following:

- *Instructor's Manual* A complimentary Instructor's Manual is available upon adoption of this sixth edition. Included in the manual are general comments about the experiments, suggested unknowns, sample quizzes, sample data tables, solutions to the postlaboratory assignments, and a master reagent list. To simplify stockroom ordering, the Instructor's Manual that accompanies the sixth edition has a list of addresses and phone numbers for suppliers for all chemicals and equipment.

- *Chemicals List* Each experiment has a designated section for chemicals and equipment. The Instructor's Manual has an alphabetical master list of all required solid chemicals, metals, solutions, organic liquid and solid compounds. The master list also indicates the experiment(s) requiring the reagent and directions for preparing solutions.

- *Laboratory Safety* Every effort has been made to ensure the safety of the student in the laboratory. General safety rules are presented in the introductory section of the manual and questions on safety appear in Experiment 1. Within each experiment students are alerted to procedures or chemicals that constitute a danger by the notation: **SAFETY**. The prelaboratory assignment of each experiment has a question about any potential safety hazards; even minor hazards such as fragile glassware are indicated in *Appendix II*.

- *Time Frame* Each experiment is designed to provide three hours of lab work, assuming ideal conditions. Since each laboratory situation is different, for example, the number and type of balances available, the time required will necessarily vary. In order to reduce the time required to perform the experiment the instructor has the flexibility of reducing the number of experimental trials or eliminating a procedure.

- *Prelaboratory Assignment* In the interest of safety and in order to provide adequate student preparation, the answers to the prelab are found in *Appendix II*. To ensure that the student is prepared for lab, the Instructor's Manual contains a sample quiz for each experiment which is based on the prelaboratory assignment.

- *Postlaboratory Assignment* The postlab questions complete the laboratory learning experience. The sixth edition postlabs have been revised and each postlab includes an optional question.

TO THE STUDENT

You are about to discover a science that has occupied the best minds of men and women since the times of the early Chinese, Egyptian, and Greek civilizations. You will be participating in a most interesting experience designed to stimulate your ability to make observations and draw logical conclusions. The first experiment, in particular, is designed to introduce the scientific method as well as to provide for the discovery of the joy of chemistry.

In order to have a positive learning experience, you should become familiar with the experiment *before* coming to laboratory. The following steps will help prepare you for a successful laboratory period:

1. Read over the **objectives** and begin to formulate the general nature of your experience.
2. Follow the **discussion** carefully, especially any calculations that will be required. If the discussion is too brief or unclear to you, refer to the appropriate section of your book.
3. Read the **equipment and chemicals** section. Refer to the illustrations in Common Laboratory Equipment for any items that are not familiar to you.
4. Outline the **procedure** so that you begin to anticipate the flow of activity during the experiment.
5. Do the **prelaboratory assignment** and check your answers in *Appendix II*.
6. Before coming to laboratory, check to make sure you have your laboratory manual, calculator, and other materials that may be required by your instructor.
7. During the experiment, record all observations directly in the **data table** or notebook as directed by your instructor. Do not record data on scratch paper.
8. During the experiment, follow these work instructions:
   (a) Be aware of the **Safety Precautions**, page 1.
   (b) Work independently unless instructed otherwise.
   (c) Pour excess liquid reagents into the sink and wash away with water.
   (d) Place excess solid reagents in a ceramic crock or other waste container.
   (e) Dispose of organic chemicals in the special waste container provided by the instructor.
   (f) When the experiment calls for water, use distilled water. When cleaning glassware, use tap water and then rinse with distilled water.
   (g) Never place chemicals directly on the balance pan; clean up spilled chemicals immediately.
   (h) Clean your laboratory station after finishing the experiment.
9. On completion of the experiment, check the **data table** to see that all measurements are recorded correctly. Check all calculations for the correct significant digits. Check formulas of compounds and make sure ions have the proper charge. All chemical equations should be correctly balanced.

10. After checking the data table for accuracy, complete the postlaboratory assignment. This exercise is intended to synthesize the learning experience. Although the postlab is related to the experiment, these questions are somewhat more general in nature.

If you follow these ten simple guidelines you should have a meaningful and worthwhile laboratory experience. In addition, the theoretical principles presented in lecture will come into focus and seem much more relevant.

Charles H. Corwin
Department of Chemistry
American River College
Sacramento, CA 95841

LABORATORY EXPERIMENTS

Seventh Edition

BASIC CHEMISTRY

Safety Precautions

The laboratory can be but is not necessarily a dangerous place. With intelligent precautions and a proper understanding of techniques, the laboratory is no more dangerous than any other classroom. Most of the precautions are just common sense practices.

1. Wear safety glasses or goggles at all times while working in the laboratory.
2. Wear shoes at all times.
3. Eating, drinking, and smoking are prohibited in the laboratory at all times.
4. Know where to find and how to use the first-aid equipment and fire extinguisher.
5. Consider all chemicals to be hazardous unless instructed otherwise.
6. If chemicals come into contact with your skin or eyes, wash immediately with large amounts of water and then consult your laboratory instructor.
7. Never directly smell any vapor or gas. Instead waft a small sample toward your nose.

Waft toward your nose.

8. Any reactions involving dangerous chemicals or unpleasant odors are to be performed in the hood.

9. Never point a test tube which you are heating at yourself or your neighbor—it may erupt like a geyser.

10. Always pour acids into water, not water into acid, because the heat of solution will cause the water to boil and the acid to splatter.
11. When inserting glass tubing or thermometers into stoppers, *lubricate the tubing and the hole in the stopper with glycerol or water.* Wrap the glass in a towel and grasp the tubing as close to the end being inserted as possible. Slide the glass into the rubber stopper with a twisting motion. Keep your hands as close together as possible in order to eliminate the possibility of breakage.

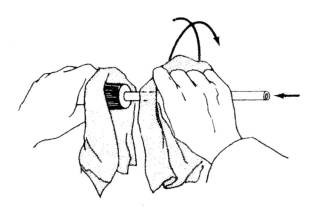

12. Clean up all broken glassware *immediately.*
13. Many common reagents, for example, alcohols, acetone, and especially ether, are highly flammable. *Do not use them anywhere near open flames.*
14. No unauthorized experiments are to be performed.
15. Observe all special precautions mentioned in the Prelaboratory Assignment of each experiment.
16. NOTIFY THE INSTRUCTOR IMMEDIATELY IN CASE OF AN ACCIDENT.

Suggested Locker Inventory

EQUIPMENT	QUANTITY
beakers, 100, 150, 250, 400, 600, 1000 mL	(6)
clay triangle	(1)
crucible and cover	(1)
crucible tongs	(1)
dropper pipet	(1)
evaporating dish	(1)
flasks, 125 mL Erlenmeyer	(3)
250 mL Erlenmeyer	(3)
1000 mL Florence	(1)
funnel	(1)
graduated cylinder, 100 mL	(1)
litmus paper, red and blue	(2)
rubber policeman	(1)
stirring rods, thin glass	(2)
test tubes, 16 × 150 mm	(6)
test tubes, 13 × 100 mm	(3)
test tube brush	(1)
test tube holder	(1)
test tube rack	(1)
thermometer, 110°C	(1)
wash bottle, plastic	(1)
watchglass, about 100 mm	(1)
wire gauze	(1)

SECTION _____   NAME _____

ASSIGNED LOCKER # _____

COMBINATION _____

COMMON LABORATORY EQUIPMENT

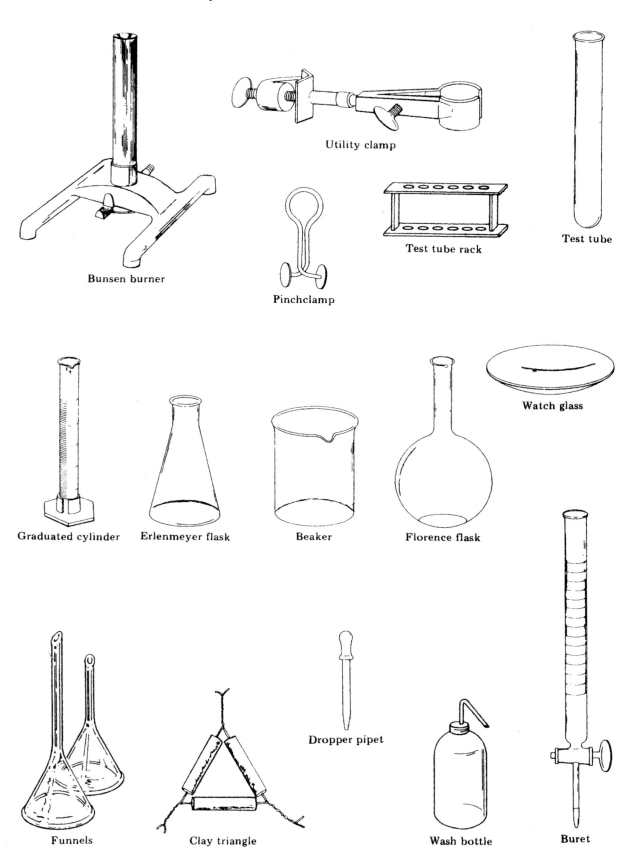

Bunsen burner

Utility clamp

Pinchclamp

Test tube rack

Test tube

Graduated cylinder

Erlenmeyer flask

Beaker

Florence flask

Watch glass

Funnels

Clay triangle

Dropper pipet

Wash bottle

Buret

COMMON LABORATORY EQUIPMENT

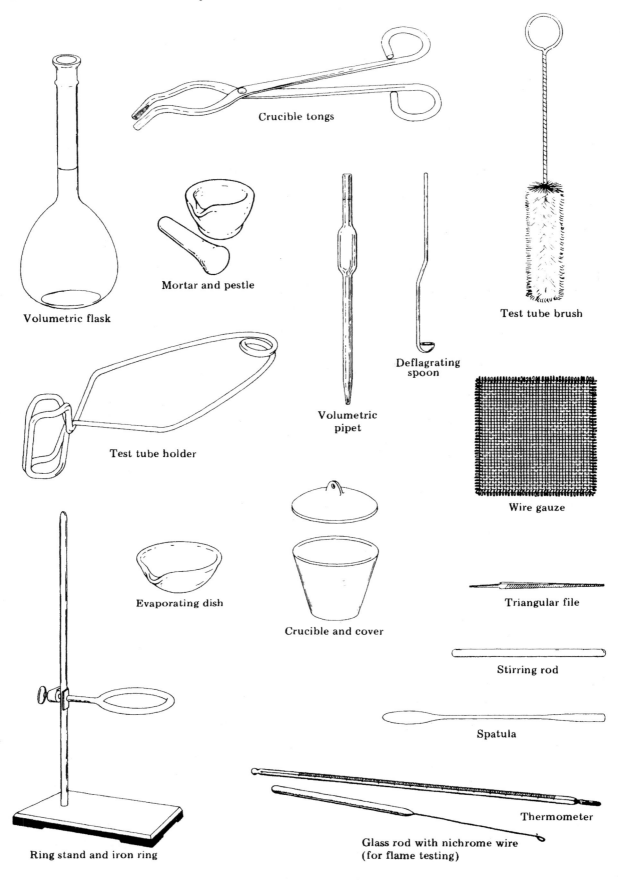

Crucible tongs

Volumetric flask

Mortar and pestle

Volumetric pipet

Deflagrating spoon

Test tube brush

Test tube holder

Wire gauze

Ring stand and iron ring

Evaporating dish

Crucible and cover

Triangular file

Stirring rod

Spatula

Thermometer

Glass rod with nichrome wire (for flame testing)

1

Scientific Observations

OBJECTIVES

1. To gain experience in recording and explaining experimental observations.
2. To develop skill in handling glassware and transferring solid and liquid chemicals.
3. To become familiar with safety precautions in the laboratory.

DISCUSSION

Chemistry is the branch of science that studies matter and the changes that matter undergoes. Science can be defined simply as organized knowledge. Scientific knowledge is gathered systematically by performing thoughtful experiments, carefully recording observations, and ultimately drawing some conclusions. This procedure is known as the scientific method and involves three steps (Figure 1-1).

1. Experimentation –collecting data by observation of chemical changes under controlled conditions.
2. Hypothesizing – formulating a tentative proposal to correlate and explain the experimental data.
3. Theorizing – stating a formal theory or scientific law after extensive testing of the hypothesis.

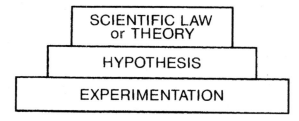

FIGURE 1-1   The scientific method.

Hypotheses are frequently proven invalid although not always immediately. Historically, chemists and physicists have been slow to abandon an acceptable theory in order to adopt a new one. Scientists exercise caution in drawing conclusions, knowing that nature reveals itself in glimpses and at times appears contradictory. Hypotheses may be discarded, modified, or on rare occasions, after rigorous testing, be elevated to the status of a scientific law or theory.

PROBLEM EXAMPLE 1-1

Mercury oxide, an orange powder, is placed in a test tube and heated for two minutes. A wooden splint is ignited and extinguished. The glowing splint is then inserted into the test tube.

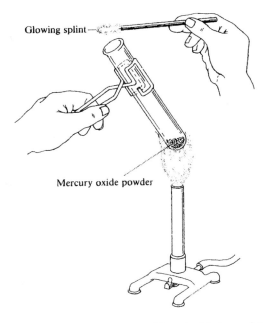

Glowing splint

Mercury oxide powder

FIGURE 1-2 Heating a compound in a test tube and testing for an evolved gas.

Observation

- A silver metal forms on the inside of the test tube.
- The glowing splint bursts into flames.

Hypothesis

- Mercury and oxygen are produced from heating mercury oxide.

EQUIPMENT AND CHEMICALS

A. Instructor Demonstrations

Equipment

- 1000 mL Florence flask with stopper
- large Erlenmeyer flask with stopper
- stir rod
- 150 mL beaker
- matches
- fire extinguisher
- mortar and pestle
- wash bottle
- evaporating dish

Chemicals

- disappearing blue solution (10 g glucose in 300 mL 0.5 M KOH + 10 mL of 0.1 g/L methylene blue solution)
- copper, Cu metal
- concentrated nitric acid, 16 M HNO_3
- sugar, powdered $C_{12}H_{22}O_{11}$
- concentrated sulfuric acid, 18 M H_2SO_4
- ethanol, C_2H_5OH
- ammonium nitrate, solid NH_4NO_3
- zinc, Zn powder

B. Student Experiments

- 16 × 150 mm test tubes (2)
- scoopula
- 250 mL beaker
- graduated cylinder
- ball-and-stick models

- ammonium chloride, solid NH_4Cl
- calcium chloride, solid $CaCl_2$
- iron metal (e.g., a nail)
- calcium, Ca metal
- copper (II) sulfate solution, 0.1 M $CuSO_4$
- mercury (II) nitrate solution, 0.1 M $Hg(NO_3)_2$
- potassium iodide solution, 0.1 M KI

PROCEDURE

A. Instructor Demonstrations. The following experiments are intended to provide interesting chemical demonstrations. The instructor while performing the demonstrations may wish to discuss laboratory safety.

You will record your observations in the Data Table and then propose a hypothesis to explain the observed event.

1. *Disappearing Blue.* Observe the clear solution in the stoppered 1000 mL Florence flask. Lift the flask and shake it once with your thumb firmly on the stopper. Repeat the procedure a couple of times.

2. *Copper Smog.* Place a piece of copper metal (for example, a penny) in a large Erlenmeyer flask. Pour concentrated nitric acid into the flask so as to cover the metal and stopper tightly.

   NOTE: After the metal has stopped reacting, the instructor may wish to empty the flask contents into a beaker of water.

3. *Foam.* Half fill a 150 mL beaker with powdered sugar. Add 15 mL of concen- d sulfuric acid and stir slowly with a glass rod.

4. *Cold Heat.* Add 40 mL of ethanol to 60 mL of water in a 150 mL beaker. Solicit a clean handkerchief from a student and soak it in the alcohol solution. Squeeze the excess solution out of the handkerchief, spread it on a lab bench, and ignite it.

   NOTE: The effect is better with the lights dimmed. The instructor may wish to use the fire extinguisher and discuss the flammability of alcohol.

5. *Water Hazard.* Grind about 3 g of ammonium nitrate in a mortar with a pestle. Empty the powder into an evaporating dish. Liberally sprinkle fresh zinc dust over the mixture. Stand back and play a stream of distilled water from a wash bottle onto the chemicals.

   NOTE: The reaction is quite exothermic and should be performed with *caution*. A few crystals of iodine enhance the effect.

B. Student Experiments. Record your observations for each of the following in the Data Table. Propose hypotheses to explain your observations.

1. *Hot and Cold.* Add a scoopula of ammonium chloride to one test tube and calcium chloride to the other. Half fill each test tube with distilled water. Place your hand around the bottom of each test tube.

   NOTE: Empty chemicals into the sink followed by water.

2. *Active and Unreactive.* Half fill a 250 mL beaker with distilled water. Place an iron nail and a piece of calcium metal in the water. Record your observations and make a hypothesis.

3. *Copper Nails.* Half fill a 250 mL beaker with copper(II) sulfate solution. Place an iron nail in the solution. Wait a few minutes then record your observations.

4. *Here and Gone.* Measure about 10 mL of mercury(II) nitrate into a graduated cylinder. Add 20 mL of potassium iodide solution into the graduated cylinder. Record your observations.

   Add an additional 30 mL of potassium iodide into the graduated cylinder and mix the contents. Record your observations and formulate an hypothesis.

5. *Mirror Images.* Given a ball-and-stick model kit, construct the model shown in Figure 1-3. The abbreviations below are as follows: B—black, Y—yellow, O—orange, R—red, and G—green.

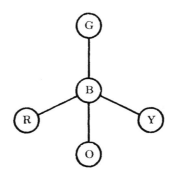

FIGURE 1-3 Ball-and-stick model.

Using additional balls and sticks, construct a second model identical to the first. Notice the two models are superimposable.

On one of the models, exchange the positions of the red and yellow after disconnecting the balls from the sticks. Are the two models now superimposable? Diagram each model in the Data Table.

PRELABORATORY ASSIGNMENT*

1. In your own words define the following terms: chemistry, experimentation, hypothesis, science, scientific law, scientific method, theory.

2. Identify the following: Erlenmeyer flask, beaker, wash bottle, Florence flask, test tube, graduated cylinder, mortar and pestle. See Common Laboratory Equipment, pages 4 and 5.

3. Where are directions given for transferring a solid or a liquid from a reagent bottle?

4. Which of the following chemicals should be handled knowledgeably and carefully: concentrated acids, alcohol, ammonium nitrate, calcium chloride, organic chemicals, distilled water?

5. What should you do if any chemical comes in contact with your skin?

*Answers in Appendix II.

DATA TABLE FOR SCIENTIFIC OBSERVATIONS

A. Instructor Demonstrations

1. *Disappearing Blue*

   Observation   *Hypothesis*

2. *Copper Smog*

   Observation   *Hypothesis*

3. *Black Foam*

   Observation   *Hypothesis*

4. *Cold Heat*

   Observation   *Hypothesis*

5. *Water Hazard*

   Observation   *Hypothesis*

B. Student Experiments

1. Hot and Cold

Observation   *Hypothesis*

2. Active and Unreactive

Observation   *Hypothesis*

3. Copper Nails

Observation   *Hypothesis*

4. Here and Gone

Observation   *Hypothesis*

5. Mirror Images

Observation   *Hypothesis*

1. The principles of chemistry have been developed by using the scientific method. State and describe the three steps of this procedure.

    (a)

    (b)

    (c)

2. Indicate in the space provided whether each of the following laboratory safety precautions are *true* or *false*.

    (a) _____ Wear safety glasses or goggles at all times while working in the laboratory.

    (b) _____ Wear shoes at all times.

    (c) _____ Eating, drinking, and smoking are strictly prohibited in the laboratory at all times.

    (d) _____ Know where to find and how to use safety and first-aid equipment.

    (e) _____ Consider all chemicals to be hazardous unless instructed otherwise.

    (f) _____ If chemicals come into contact with your skin or eyes, wash immediately with copious amounts of water and then consult your laboratory instructor.

    (g) _____ Never taste anything. Never directly smell the source of any vapor or gas; instead, by means of your cupped hand, bring a small sample to your nose.

    (h) _____ Any reactions involving skin-irritating or dangerous chemicals, or unpleasant odors, are to be performed in the hood.

    (i) _____ When heating a chemical in a test tube, never point the open end toward yourself or your neighbor.

    (j) _____ No unauthorized experiments are to be performed.

    (k) _____ Clean up all broken glassware immediately.

    (l) _____ Always pour acids into water, not water into acid, because the heat of solution will cause the water to boil and the acid to splatter.

    (m) _____ When inserting glass tubing or thermometers into stoppers, lubricate the tubing and the hole in the stopper with glycerol or water.

    (n) _____ Do not use alcohol, acetone, or ether near open flames.

    (o) _____ Observe all special precautions mentioned in each experiment.

    (p) _____ Notify the instructor immediately in case of an accident.

3. State whether the following instructions for working in the laboratory are *true* or *false*.

(a) _____ Read the experiment before coming to the laboratory: the Objectives, Discussion, Equipment and Chemicals, and Procedure. Do the Prelaboratory Assignment and check your answers in Appendix II.

(b) _____ Work independently unless instructed otherwise.

(c) _____ Record your observations directly in your Data Table or notebook. Do not record data on scratch paper.

(d) _____ Be aware of safety precautions and avoid accidents.

(e) _____ Pour excess liquid reagents into the sink and wash away with water.

(f) _____ Place excess solid reagents in the crock or other waste container.

(g) _____ Dispose of organic chemicals in the special waste container provided by the instructor.

(h) _____ When the experiment calls for water, use distilled water. When cleaning glassware, use tap water and then rinse with distilled water.

(i) _____ Never place chemicals directly on the balance pan.

(j) _____ Never place a hot or warm object on the balance pan. Allow the object to cool to room temperature before weighing.

(k) _____ After heating an object, do not place it on the desk top. Allow it to cool or place it on a wire gauze.

(l) _____ Clean your laboratory station upon completion of the experiment.

4. (optional) You are given nine pennies and a platform balance.

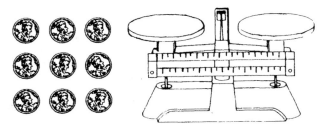

One penny is lighter than the other eight of equal mass. How can you determine the light penny in only two weighings?

Metric System Measurements

OBJECTIVES

1. To obtain measurements of mass, length, volume, temperature, and time.
2. To determine the mass and calculate the volume of an unknown rectangular solid.
3. To gain proficiency in using the following instruments: balances, metric rulers, graduated cylinder, thermometer, clock.

DISCUSSION

Beginning in France in 1790, a standard system of measurement for length, mass, and volume was proposed. The system was called the metric system after the French word *metrè*, meaning "to measure" (Table 2-1). In addition to utilizing a basic set of units, the system had the advantage of using prefixes. Prefixes made the basic units larger or smaller by multiples or fractions of 10. That is, the proposed system was a decimal system. For instance, one kilometer converts to 1000 meters or 100,000 centimeters. Compare this system to the English system, in which one mile converts to 1760 yards or 63,360 inches. The English system does not use basic units, nor are the units decimally related. Despite its logic and simplicity, adoption of the metric system took place slowly, and not until the Treaty of the Meter in 1875 was there international acceptance by most major countries. Today the metric system, in one form or another, is used worldwide within the scientific community.

Scientific measurements have gradually progressed to a high state of refinement. However, it is still not possible, and never will be, to make an exact measurement, because all measurements are made with instruments that have an inherent degree of error. The amount of error is reflected in the uncertainty of the measurement. For example, a sweep-second clock measures time to the nearest second. Digital clocks commonly measure time to a tenth or a hundredth of a second. The uncertainty is ± 1 s, ± 0.1 s, and ± 0.01 s for the three clocks. The clock with the least uncertainty (± 0.01 s) gives the most precise measurements.

In this experiment, we will become familiar with several instruments. We will take mass measurements with balances having progressively greater precision. A decigram balance is so named because the finest division is one-tenth of a gram. It follows that the precision of a centigram balance is one-hundredth of a gram. A milligram balance provides mass measurements to one-thousandth of a gram.

TABLE 2-1 METRIC UNITS OF MASS, LENGTH, AND VOLUME

Quantity	Basic Unit	Multiple/Fraction of Basic Unit	Derived Unit	Symbol
mass	gram	$1000 \ (10^{3})$	kilogram	kg
		$0.1 \ (10^{-1})$	decigram	dg
		$0.01 \ (10^{-2})$	centigram	cg
		$0.001 \ (10^{-3})$	milligram	mg
length	meter	$1000 \ (10^{3})$	kilometer	km
		$0.1 \ (10^{-1})$	decimeter	dm
		$0.01 \ (10^{-2})$	centimeter	cm
		$0.001 \ (10^{-3})$	millimeter	mm
volume	liter	$1000 \ (10^{3})$	kiloliter	kL
		$0.1 \ (10^{-1})$	deciliter	dL
		$0.01 \ (10^{-2})$	centiliter	cL
		$0.001 \ (10^{-3})$	milliliter	mL

We will make **length** measurements using two different metric rulers. One ruler is calibrated in centimeter divisions; the other in tenth centimeter subdivisions. The ruler having centimeter divisions gives data with more uncertainty. The following examples demonstrate measurement of length.

PROBLEM EXAMPLE 2-1

A copper rod is measured with the metric ruler shown below. What is the length of the rod?

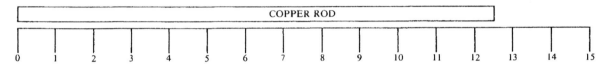

METRIC RULER A   *(Estimate to a tenth of a division ±0.1 cm)*

Solution: Each division represents one centimeter. The end of the rod lies between the 12th and 13th division. We can estimate to a tenth of a division (±0.1 cm). Since the end of the rod lies about five-tenths past 12, we can estimate the length as

$$12 \text{ cm} + 0.5 \text{ cm} = 12.5 \text{ cm}$$

PROBLEM EXAMPLE 2-2

The same copper rod is measured with the metric ruler shown below. What is the length of the rod?

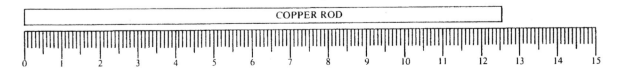

METRIC RULER B   *(Estimate to a half of a division ±0.05 cm)*

Solution: Note that this ruler is divided into centimeters which are subdivided into tenths of centimeters. The end of the rod lies between the 12th and 13th divisions, and between the 5th and 6th subdivisions. Therefore, the length is between 12.5 cm and 12.6 cm.

We can estimate the measurement more precisely. A subdivision is too small to divide into ten parts but we can estimate to half of a division (± 0.05 cm). The length is

$$12 \text{ cm} + 0.5 \text{ cm} + 0.05 \text{ cm} = 12.55 \text{ cm}$$

We can obtain the **volume** of a liquid using a graduated cylinder. If we examine the 100 mL graduated cylinder shown in Figure 2-1, we notice that every ten divisions are marked. Ten divisions equal 10 milliliters; therefore, each division represents one milliliter. If we estimate to half a division, the volume measurement has an uncertainty of ± 0.5 mL.

When reading a graduated cylinder, always read the bottom of the meniscus (lens-shaped surface of the liquid). Observe the meniscus at eye level in order to avoid a reading error.

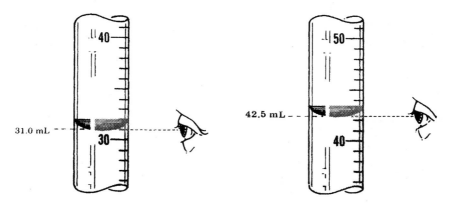

FIGURE 2-1 Example readings obtained using proper eye position and reading the bottom of the meniscus.

To measure **temperature** we will use a $110°C$ Celsius thermometer. In Figure 2-2 the thermometer is divided into ten-degree intervals and further subdivided into one-degree intervals. Thus, each division represents one degree Celsius. By estimating between divisions, we can obtain measurements with an uncertainty of $\pm 0.5°C$.

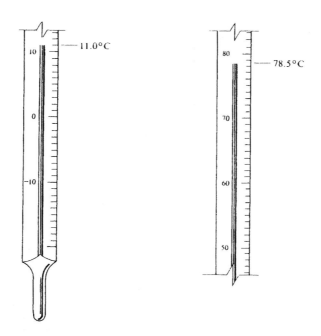

FIGURE 2-2 Example readings using a Celsius thermometer.

We can measure **time** by observing a clock or watch with a sweep-second hand. The uncertainty is one division; that is, ±1 s. Alternately, we can use a digital timepiece. Digital watches having an uncertainty of ±0.01 s or ±0.001 s are commonplace.

TABLE 2-2 MEASUREMENT OF MASS, LENGTH, VOLUME,
TEMPERATURE, AND TIME

Quantity	Instrument	Uncertainty	Example Readings
mass	decigram balance	±0.1 g	86.4 g, 101.7 g
	centigram balance	±0.01 g	86.32 g, 101.73 g
	milligram balance	±0.001 g	86.318 g, 101.730 g
length	metric ruler A	±0.1 cm	5.2 cm, 10.1 cm
	metric ruler B	±0.05 cm	5.20 cm, 10.05 cm
volume	graduated cylinder	±0.5 mL	27.5 mL, 60.0 mL
temperature	thermometer	±0.5°C	1.0°C, 23.5°C
time	sweep-second clock	±1 s	42 s, 52 s
	hundredth-second clock	±0.01 s	45.31 s, 51.95 s

To test our skill in making metric measurements we will find the mass and volume of an unknown rectangular solid. The volume of a rectangular solid is determined by multiplying its length, width, and thickness. The following examples will illustrate.

PROBLEM EXAMPLE 2-3

An unknown rectangular solid has the following dimensions: 5.0 cm by 2.5 cm by 1.1 cm. What is the volume of the solid?

Solution: The volume of a rectangular solid is equal to its length times width times thickness.

$$5.0 \text{ cm} \times 2.5 \text{ cm} \times 1.1 \text{ cm} = 13.75 \text{ cm}^3$$

In multiplication, the product is limited by the least number of significant digits. In this example each dimension has two digits. Thus, the volume is limited to two significant digits. The correct volume, after rounding off, is 14 cm^3.

NOTE: A cm^3 is abbreviated cc in the medical field, but this is discouraged in science. The volume of a liquid is often expressed in milliliters. A cubic centimeter is exactly equal to one milliliter: 1 cm^3 = 1 mL.

PROBLEM EXAMPLE 2-4

The unknown rectangular solid in the previous example was remeasured with a more precise ruler and the following dimensions were obtained: 5.00 cm by 2.45 cm by 1.15 cm. What is the volume of the solid?

Solution: The volume is once again found by multiplying together the three dimensions.

$$5.00 \text{ cm} \times 2.45 \text{ cm} \times 1.15 \text{ cm} = 14.0875 \text{ cm}^3$$

This time each dimension has three significant digits so the volume should have three. Rounding off to three significant digits, the volume becomes **14.1 cm^3**.

1-31-2013

EQUIPMENT AND CHEMICALS

- decigram balance
- centigram balance
- milligram balance
- 250 mL beaker
- crucible and cover
- watchglass
- evaporating dish
- 100 mL graduated cylinder
- dropper pipet
- 16 × 150 mm test tubes (3)

- 110°C thermometer
- wire gauze
- unknown rectangular solid
- nickel (5¢ coin)

- *Solution A*, (2 g KIO$_3$/L)
- *Solution B*, (dissolve 1 g starch in 1 L of boiling water; add 0.4 g NaHSO$_3$, 5 mL 1 M H$_2$SO$_4$)

PROCEDURE · 1/31/13

A. Mass Measurement

1. Determine the mass of a 250 mL beaker on each of the following: (a) decigram balance, (b) centigram balance, (c) milligram balance.
2. Determine the mass of a crucible and cover on each of the three balances.

NOTE: Refer to balance instructions in Appendices V, VI, and VII. The instructor may choose to omit one or more of the above weighings according to balance availability.

B. Length Measurement

1. Measure the diameter of a watchglass with each of the following: (a) metric Ruler A, (b) metric Ruler B.
2. Measure the diameter of an evaporating dish on each of the two rulers.

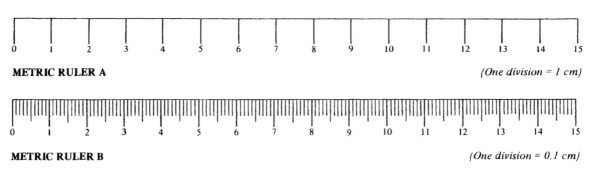

METRIC RULER A   *(One division = 1 cm)*

METRIC RULER B   *(One division = 0.1 cm)*

FIGURE 2-3 Metric Rulers

C. Volume Measurement

1. Fill a 100 mL graduated cylinder with water. Adjust the bottom of the meniscus to the full mark with a dropper pipet. Record the volume as 100.0 mL.
2. Fill a test tube with water from the graduated cylinder. Record the new meniscus reading in the cylinder (± 0.5 mL).
3. Fill a second test tube with water. Record the volume in the graduated cylinder.
4. Fill a third test tube with water. Record the volume in the graduated cylinder.

D. Temperature Measurement

1. Record the temperature in the laboratory using a Celsius thermometer ($\pm 0.5^{\circ}$C).
2. Half fill a 250 mL beaker with ice and water. Insert the thermometer into the beaker and record the coldest observed temperature.
3. Half fill the 250 mL beaker with distilled water. Support the beaker on a ring stand with a wire gauze as shown in Figure 2-4. Heat the water to boiling with a laboratory burner and record the hottest observed temperature.

   NOTE: Hold the thermometer off the bottom of the beaker to avoid an erroneously high reading.

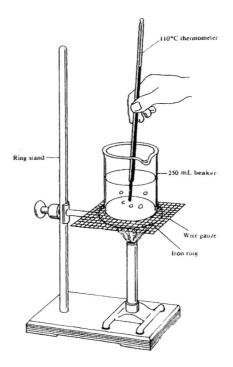

FIGURE 2-4 Waterbath for boiling water.

E. Time Measurement

1. Fill one test tube with *Solution A* and a second test tube with *Solution B*. Pour both solutions simultaneously into a 250 mL beaker. Record the number of seconds for a reaction to occur; this is indicated by a color change.

F. Mass and Volume of an Unknown Rectangular Solid

1. Obtain a rectangular solid and record the unknown number in the Data Table. Find the mass of the unknown solid using a decigram, centigram, and milligram balance.
2. Measure the length, width, and thickness of the rectangular solid unknown using Ruler A in Figure 2-3. Calculate the volume.
3. Measure the length, width, and thickness of the rectangular solid unknown using Ruler B in Figure 2-3. Calculate the volume.

G. Metric Estimations

1. Estimate the mass of a nickel. Weigh the nickel on any balance and record the mass ±1 g.
2. Estimate the diameter of a nickel. Measure the nickel with any metric ruler and record the length ±1 cm.
3. Estimate the volume of 20 drops of water. Using a dropper pipet, add 20 drops of water into a graduated cylinder and record the volume ±1 mL.

PRELABORATORY ASSIGNMENT*

1. In your own words, define the following terms: balance, mass, meniscus, metric system, parallax, uncertainty, weight.
2. Identify the following laboratory equipment from the diagrams on pages 4–5: beaker, crucible and cover, watchglass, evaporating dish, graduated cylinder, test tube, thermometer.
3. State the quantity expressed by each of the following units.

   (a) gram — g

   (b) centimeter — cm

   (c) milliliter — mL

   (d) degree Celsius — °C

   (e) second — s

4. State the uncertainty in the measurements obtained from the following metric instruments.

   (a) decigram balance

   (b) centigram balance

   (c) milligram balance

   (d) metric Ruler A

   (e) metric Ruler B

   (f) graduated cylinder

   (g) thermometer

   (h) sweep-second clock

*Answers in Appendix II.

5. Record the measurement indicated by each of the following instruments. The reading must be consistent with the uncertainty of the instrument.

(a) metric rulers

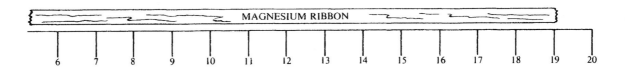

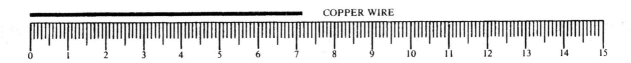

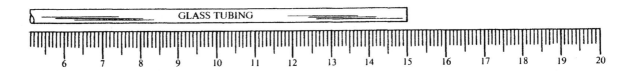

GLASS TUBING

(b) graduated cylinders

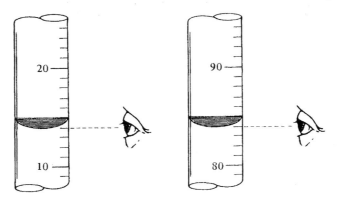

(c) thermometers

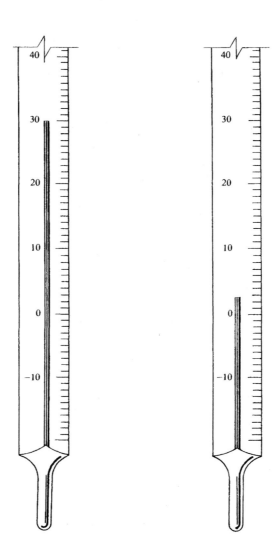

6. An unknown rectangular solid has the following measurements: 3.70 cm by 2.45 cm by 1.25 cm. Calculate the volume in cubic centimeters.

7. What safety precautions must be observed in this experiment?

Density of Liquids and Solids

OBJECTIVES

1. To determine the density of the following: water, unknown liquid, rubber stopper, unknown rectangular solid.
2. To calculate the thickness of aluminum foil given its density and the measurements of mass, length, and width.
3. To gain proficiency in performing the following experimental procedures: pipetting a liquid, weighing by difference, and determining volume by displacement.

DISCUSSION

Density is a physical property of matter that is defined as the amount of mass per unit volume; in equation form we have

$$\text{density (d)} = \frac{\text{mass (m)}}{\text{volume (V)}}$$

To determine experimentally the density of a liquid or solid, we first measure the mass using a balance. The volume can be obtained using calibrated glassware or by calculation. After collecting the data, the density is calculated from the ratio of mass to volume. In addition, the proper units must be attached to the calculated value. The density of liquids and solids is usually expressed in grams per milliliter (g/mL) or grams per cubic centimer (g/cm^3). Since by definition 1 mL = 1 cm^3, the numerical value for the density of a liquid or solid is identical in units of g/mL or g/cm^3.

To determine the mass of a liquid we will use an indirect technique called *weighing by difference* (Figure 3-1). First, we will weigh a flask empty. Next we will pipet a given volume of liquid into the flask and reweigh. The mass of the liquid is found by subtraction.

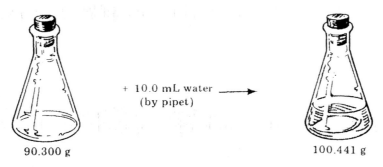

90.300 g   100.441 g

FIGURE 3-1 Weighing by difference;
100.441 − 90.300 = 10.141 g of water.

PROBLEM EXAMPLE 3-1

A 10.0 mL sample of water is pipetted into a flask. The mass of water is found from weighing by difference (see Figure 3-1). If the mass is 10.141 g, calculate the density of water.

Solution: Dividing mass by volume,

$$\frac{10.141 \text{ g}}{10.0 \text{ mL}} = 1.01 \text{ g/mL}$$

The answer is limited to three significant digits by the value in the denominator. The calculated value, 1.01 g/mL, agrees well with the theoretical value, 1.00 g/mL. The slight discrepancy is due to experimental error.

The volume of an irregular object cannot be found directly. However, its volume can be found indirectly from the amount of water it displaces. This technique is called *volume by displacement*. For example, the volume of a rubber stopper can be determined as shown in Figure 3-2. The initial reading of water in the graduated cylinder is recorded. The stopper is introduced into the cylinder and then the final reading is taken. The difference between the initial and final readings corresponds to the volume of water displaced. The volume of water displaced is equal to the volume of the rubber stopper.

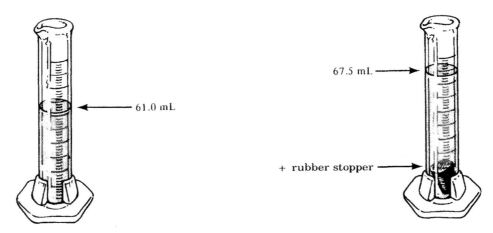

FIGURE 3-2 Volume by displacement for a rubber stopper;
67.5 − 61.0 = 6.5 mL.

PROBLEM EXAMPLE 3-2

A rubber stopper weighing 8.453 g displaces 6.5 mL of water in a graduated cylinder (Figure 3-2). What is the density of the stopper?

Solution: Dividing mass by volume,

$$\frac{8.453 \text{ g}}{6.5 \text{ mL}} = 1.3 \text{ g/mL}$$

In this example the volume has only two digits. Thus, the density is limited to two significant digits.

The volume of a solid object with regular dimensions can be found by calculation. For example, the volume of a rectangular solid is equal to its length times width times thickness.

PROBLEM EXAMPLE 3-3

The mass of a rectangular solid stainless steel block is 139.443 g. If the block measures 5.00 cm by 2.55 cm by 1.25 cm, what is the density of the stainless steel?

Solution: First, we calculate the volume of the block:

$$5.00 \text{ cm} \times 2.55 \text{ cm} \times 1.25 \text{ cm} = 15.9 \text{ cm}^3$$

Second, we find the density of the stainless steel:

$$\frac{139.443 \text{ g}}{15.9 \text{ cm}^3} = 8.77 \text{ g/cm}^3$$

The thickness of a sheet of metal foil is too thin to measure with a ruler. However, we can find the thickness indirectly by calculation. Given the mass, length, and width of a metal foil, we can use the density of the metal to calculate the thickness of the foil.

PROBLEM EXAMPLE 3-4

A rectangular sheet of tin foil has a mass of 0.571 g and measures 5.10 cm by 10.25 cm. Given the density of tin, 7.28 g/cm^3, calculate the thickness of the foil.

Solution: To calculate the thickness of the foil we must first find the volume of the foil. The volume can be calculated from the density of tin as follows:

$$0.571 \text{ g} \times \frac{1 \text{ cm}^3}{7.28 \text{ g}} = 0.0784 \text{ cm}^3$$

The thickness of the rectangular sheet of foil is found from dividing the volume by its length and width:

$$\frac{0.0784 \text{ cm}^3}{5.10 \text{ cm} \times 10.25 \text{ cm}} = 0.00150 \text{ cm} (1.50 \times 10^{-3} \text{ cm})$$

EQUIPMENT AND CHEMICALS

A. Instructor Demonstration

- tall glass cylinder
- methylene chloride
- hexane

- glass object
- *hard* plastic object
- ice
- cork

Student Experiments

- 125 mL Erlenmeyer flask
- 150 mL beaker
- 10 mL pipet
- pipet bulb
- unknown liquids

- 100 mL beaker
- 100 mL graduated cylinder
- #2 rubber stopper
- unknown rectangular solid
- aluminum foil, ~5 × 10 cm rectangle

PROCEDURE

 A. Density Observations — Instructor Demonstration

98.20 1. Half fill a tall glass cylinder with water. Add methylene chloride until two layers are observed. Add hexane until three layers are observed. Record the positions of each layer in the Data Table.
2. Drop a glass object into the cylinder and record the observation.
3. Drop a plastic object into the cylinder; record the observation.
4. Drop a piece of ice into the cylinder; record your observation.
5. Drop a cork into the cylinder; record your observation.

B. Density of Water

98.26 1. Weigh a 125 mL Erlenmeyer flask fitted with a rubber stopper.
2. Half fill a 150 mL beaker with distilled water and pipet 10.0 mL into the flask (see Appendix VIII).
107.70 3. Reweigh the flask and stopper and determine the mass of water by difference.
4. Repeat a second trial.

   NOTE: It is not necessary to dry the flask between trials because the 10.0 mL sample of water is weighed by difference.

5. Calculate the density of the water for each trial and report the average value for both trials.

C. Density of an Unknown Liquid

1. Obtain about 25 mL of an unknown liquid in a 100 mL beaker. Record the unknown number in the Data Table.
2. Weigh a 125 mL Erlenmeyer flask fitted with a rubber stopper.
3. Condition the pipet and transfer 10.0 mL of unknown liquid into the flask.
4. Reweigh the flask and stopper and determine the mass of liquid by difference.
5. Repeat a second trial.
6. Calculate the density of the liquid for each trial and report the average value for both trials.

D. Density of a Rubber Stopper

1. Weigh a dry #2 rubber stopper.
2. Fill a 100 mL graduated cylinder about half full with water. Record the water level by observing the bottom of the meniscus and estimating to ±0.5 mL.
3. Tilt the graduated cylinder and let the stopper slowly slide into the water. Record the new level and calculate the volume by displacement for the stopper.
4. Repeat a second trial.
5. Calculate the density of the rubber stopper for each trial and report the average value for both trials.

E. **Density of an Unknown Rectangular Solid**

1. Obtain an unknown rectangular solid and record the unknown number in the Data Table.
2. Weigh the unknown solid and record the mass.
3. Measure the length, width, and thickness with a metric ruler (see page 34). Record the data and find the volume of the rectangular solid.
4. Calculate the volume of the unknown rectangular solid.

F. **Thickness of Aluminum Foil**

1. Obtain a rectangular piece of aluminum foil.
2. Record the length and width of the foil (see page 34) in the Data Table.
3. Fold the foil twice. Weigh and record its mass.
4. Calculate the volume and thickness of the aluminum foil. (The density of aluminum is 2.70 g/cm³.)

PRELABORATORY ASSIGNMENT*

1. In your own words, define the following terms: conditioning, density, meniscus, volume by displacement, weighing by difference.
2. Record the measurement indicated by each of the following instruments. The reading should be consistent with the uncertainty of the instrument:
   (a) graduated cylinder

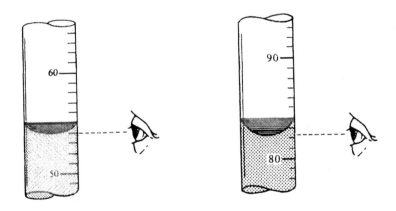

(b) metric ruler

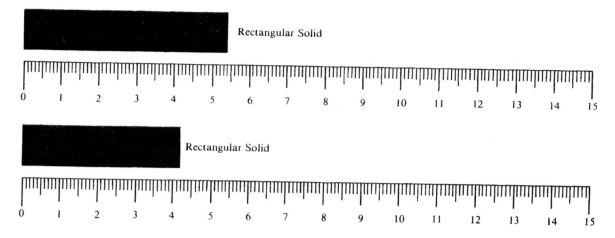

3. A 10.0 mL sample of acetone is pipetted into a stoppered flask. The mass is found by difference to be 7.899 g. Calculate the density of the liquid.

4. A piece of chalk weighing 15.60 g displaces 6.5 mL of water in a graduated cylinder. Calculate the density of the chalk.

5. A rectangular block of jade has a mass of 146.25 g and measures 10.00 cm by 3.00 cm by 1.50 cm. What is the density of the jade?

6. Find the thickness of a piece of gold foil that has a mass of 1.000 g and measures 5.00 cm by 10.00 cm. The density of gold is 18.9 g/cm^3.

7. What safety precautions must be observed in this experiment?

Specific Heat of a Metal

OBJECTIVES

1. To determine the specific heat of metallic aluminum.
2. To determine the specific heat of an unknown metal.
3. To gain practical experience in measuring temperature changes in a calorimeter.

DISCUSSION

One of the properties of matter is that it requires a certain amount of heat energy to raise the temperature of a unit mass of a substance. This property is termed the *specific heat* of a substance. In the metric system the specific heat of a substance is the amount of heat energy necessary to raise 1.00 g of that substance 1.00°C.

Heat energy is measured in units of calories (cal) or kilocalories (kcal). A calorie is the amount of heat necessary to raise the temperature of 1.00 g of water 1.00°C. Thus, we say that the specific heat of water is 1.00 calorie per gram per degree Celsius.

$$\text{specific heat of water} = \frac{1.00 \text{ cal}}{1 \text{ g} \times 1°\text{C}} \text{ or } 1.00 \text{ cal/g} \times °\text{C}$$

One of the properties of metals is that they are good conductors of heat. It therefore follows that metals have low specific heats as it requires less heat to raise their temperatures.

In this experiment we wish to find the specific heat of a metal. If a heated sample of hot metal is dropped into a styrofoam cup containing cool water, the temperature of the metal decreases and the temperature of the water increases. As a matter of fact, if we assume that no heat escapes from the cup, then the heat loss of the metal is equal to the heat gain by the water.

$$\text{heat loss of metal} = \text{heat gain of water}$$

From the temperature changes of the metal and water, we can calculate the specific heat of the metal.

PROBLEM EXAMPLE 4-1

A 50.05 g sample of zinc metal at 99.5°C was dropped into a styrofoam cup containing 100.0 g of water at 21.0°C. The water in the cup reached a maximum temperature of 24.5°. Calculate the specific heat of the metal.

Solution: First, let us calculate the heat gain of the water.

$$\frac{1.00 \text{ cal}}{1 \text{ g} \times 1°C} \times 100.0 \text{ g} \times (24.5 - 21.0)°C = 350 \text{ cal}$$

Since the temperature change, 3.5°C, represents only two significant digits, the final answer is limited to two significant digits.

The heat gain of the water is 350 calories; hence, the heat loss of the metal must be 350 calories (assuming no other heat losses to the environment). The heat loss for the sample over the temperature drop interval is

$$\frac{350 \text{ cal}}{50.05 \text{ g} \times (99.5 - 24.5)°C} = \frac{350 \text{ cal}}{50.05 \text{ g} \times 75.0°C} = 0.093 \text{ cal/g} \times °C$$

Experimentally, the metal will be first heated in a test tube immersed in a boiling waterbath; then the metal will be transferred into a styrofoam cup containing water at room temperature. Calorimeter is the technical term for a device where heat changes are measured. The styrofoam cup calorimeter and waterbath apparatus are shown in Figure 4-1.

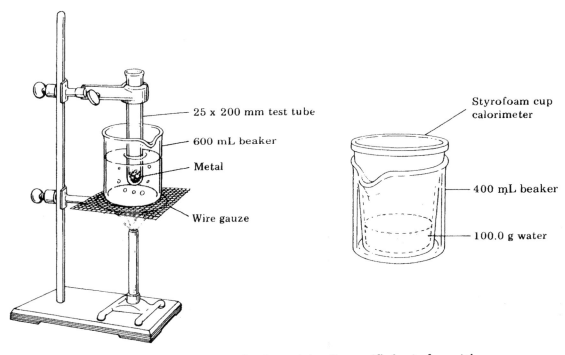

FIGURE 4-1 Apparatus for determining the specific heat of a metal.

EQUIPMENT AND CHEMICALS

- 250 mL beaker
- wire gauze
- utility clamp
- 25 × 200 mm test tube
- 600 mL beaker
- 110°C thermometer

- 100 mL graduated cylinder
- 400 mL beaker
- styrofoam cup
- aluminum rods or sticks (2 in. by 1/4 in.)
- unknown metal samples

PROCEDURE

A. Specific Heat of Aluminum

1. Place a 600 mL beaker on a wire gauze supported on a ring stand. Add about 300 mL of distilled water and bring to a boil.
2. Obtain the mass of a 250 mL beaker. Place about 50 g of aluminum rods into the beaker and weigh accurately.
3. Pour the metal rods into a 25 × 200 mm test tube and insert the test tube, suspended by a utility clamp, into the beaker of boiling water. Continue to boil the water while heating the metal (see Figure 4-1).
4. Place a styrofoam cup into a 400 mL beaker for support. From a graduated cylinder, pour 100.0 mL of water into the cup. Record the *mass* of the water.

   NOTE: Recall that the density of water is 1.00 g/mL. The mass of 100.0 mL of water would of course be 100.0 g.

5. Record the temperature of the water in the styrofoam cup.
6. Observe the temperature of the boiling water in the waterbath and record the temperature of the metal.

   NOTE: Assume the temperature of the boiling water to be the same as that of the metal. Since the thermometer is not calibrated, we cannot assume a temperature of precisely 100.0°C for boiling water.

7. Rapidly transfer the metal into the calorimeter. Stir the water in the calorimeter with the thermometer. Observe the temperature rise for several minutes and then record the maximum temperature.
8. Replace the calorimeter water, weigh out dry metal rods, and do a second trial.
9. Calculate the specific heat for each trial and the average value of both trials for the aluminum metal.

   NOTE: Dry the aluminum rods on a paper towel and return to the designated container.

B. Specific Heat of an Unknown Metal

1. Obtain an unknown metal sample from the instructor and record the unknown number.
2. Follow the previous procedure and determine the specific heat of the unknown metal.

   NOTE: After completing the experiment, dry and return the unknown metal sample to the designated container.

PRELABORATORY ASSIGNMENT*

1. In your own words define the following terms: calorie, calorimeter, specific heat.
2. What is the mass of 100.0 mL of water?
3. What is the value of the specific heat of water?
4. If the temperature change for the calorimeter water is less than ten degrees, how many significant digits are in the calculated value for specific heat?
5. A 55.50 g sample of aluminum at 99.0°C is placed in 100.0 g of water at 19.5°C, producing a maximum resulting temperature of 27.5°C. Calculate the specific heat of the aluminum sample.
6. What are the primary sources of error in this experiment?
7. What precautions must be observed in this experiment?

*Answers in Appendix II.

EXPERIMENT 4 NAME ____________

DATE ________ SECTION ____________

DATA TABLE FOR SPECIFIC HEAT OF A METAL

A. Specific Heat of Aluminum

mass of beaker + metal	________g	________g
mass of beaker	________g	________g
mass of metal	________g	________g
mass of water	________g	________g
temperature of metal	________°C	________°C
initial temperature of calorimeter water	________°C	________°C
maximum temperature of calorimeter water	________°C	________°C

Show the calculation for the specific heat of metal for trial 1.

Specific heat of metal	________cal/g × °C	________cal/g × °C

Average specific heat of metal ________cal/g × °C

B. Specific Heat of an Unknown Metal UNKNOWN #_____

mass of beaker + metal	_____ g	_____ g
mass of beaker	_____ g	_____ g
mass of unknown metal	_____ g	_____ g
mass of water	_____ g	_____ g
temperature of unknown metal	_____ °C	_____ °C
initial temperature of calorimeter water	_____ °C	_____ °C
maximum temperature of calorimeter water	_____ °C	_____ °C

Show the calculation for specific heat for trial 1.

Specific heat of unknown metal	_____ cal/g × °C	_____ cal/g × °C
Average specific heat of unknown metal	_____ cal/g × °C	

POSTLABORATORY ASSIGNMENT NAME ______________________

1. Calculate the number of kilocalories necessary to raise $25\bar{0}$ g of water from 19.6 to 68.8°C.

____

2. Calculate the mass of a piece of copper that released 125 calories when cooled from 100.0°C to 25.3°C. The specific heat of copper is 0.00924 cal/g × °C.

____

3. Find the specific heat of an unknown metal if a 35.5 g sample at 99.6°C produced a resulting temperature of 26.1°C when placed in a calorimeter containing 100.5 g of water at 20.2°C.

____

4. A 65.0 g sample of zinc (specific heat = 0.0922 cal/g × °C) was cooled from 100.0°C to 29.4°C in a calorimeter cup containing water initially at 21.5°C. Find the mass of water in the cup.

_____

5. (optional) Ignoring any heat loss to the surrounding environment, calculate the theoretical maximum temperature produced in a calorimeter containing 95.5 g of water at 20.8°C after a 25.0 g sample of aluminum (specific heat = 0.215 cal/g × °C) at 98.6°C has been introduced.

_____

Physical and Chemical Properties

Missed this Lab on Feb 14th Excused for STEPHEN

OBJECTIVES

1. To observe several metals and nonmetals.
2. To determine the boiling points of methanol and an unknown liquid.
3. To determine whether a solid substance is soluble or insoluble in water.
4. To determine whether a liquid is miscible or immiscible with water.
5. To determine whether a substance is undergoing a physical or chemical change.

DISCUSSION

One of the ways we classify matter is by its physical or chemical properties. These properties can be divided into two classes: homogeneous and heterogeneous. Homogeneous matter has consistent physical and chemical properties throughout, regardless of the size of the sample being considered. Heterogeneous matter is not consistent in terms of properties. Sugar and salt mixed together may appear to be homogeneous. Actually they are a heterogeneous mixture because their individual physical and chemical properties are quite different. Figure 5-1 illustrates the classification of matter.

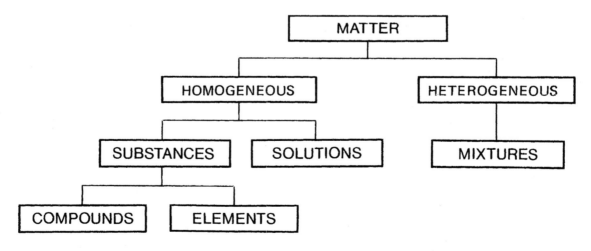

FIGURE 5-1   Classification of matter.

Sugar is a compound which is a pure substance. Upon heating, sugar decomposes into carbon and water. Using electric current, water decomposes further into hydrogen and oxygen. Both elements are colorless gases under normal conditions, but they differ in most of their other physical and chemical properties. Different elements may have some properties which are similar, but no two elements have all properties which are identical.

Physical properties are the characteristics of a substance that can be measured or observed without the substance undergoing a change in composition. The list of physical properties for a substance is quite extensive. The properties that are usually considered important include physical state (solid, liquid, gas), color, density, malleability, ductility, hardness, crystalline form, melting point, boiling point, electrical and heat conductivity, solubility, miscibility, and a few others. Odor and taste are classified as physical properties although a chemical reaction must occur in order to observe them.

The chemical properties of matter can only be observed when a substance undergoes a change in composition. That is, a chemical reaction must take place. Gasoline burning, copper metal turning green, and baking soda fizzing in acid are all examples of a chemical change. For every chemical change there is a change in energy. If heat energy is released, the substance feels hot. If heat is absorbed, it feels cool. In this experiment, we will record a chemical change if a change in color or odor is noted, a gas is released, or an insoluble substance is formed after mixing two solutions.

EQUIPMENT AND CHEMICALS

Equipment

- wire gauze
- 400 mL beaker
- 110°C thermometer
- split cork
- boiling chip
- 16 × 150 mm test tubes (6)
- 400 mL beaker
- 110°C thermometer
- split cork
- boiling chip
- 16 × 150 mm test tubes (6)
- evaporating dish
- 250 mL beaker
- test tube holder
- test tube brush
- wash bottle

Chemicals

- small vials containing samples of cobalt, hydrogen, lead, magnesium, manganese, neon, oxygen, sulfur, tin, zinc
- methyl alcohol, CH_3OH
- boiling point unknowns
- iodine, solid crystals I_2
- sucrose, solid crystals $C_{12}H_{22}O_{11}$
- amyl alcohol, $C_5H_{11}OH$
- copper wire, heavy gauge Cu
- ammonium bicarbonate, solid NH_4HCO_3
- potassium bicarbonate, solid $KHCO_3$
- sodium carbonate solution, 0.5 M Na_2CO_3
- sodium sulfate solution, 0.1 M Na_2SO_4
- dilute hydrochloric acid, 6 M HCl
- sodium nitrate solution, 0.1 M $NaNO_3$
- lead nitrate solution, 0.1 M $Pb(NO_3)_2$
- potassium iodide solution, 0.1 M KI

PROCEDURE

A. Study of Physical Properties

1. Observing Physical Properties. Examine vials containing the following elements and complete the Data Table.

   (a) cobalt
   (c) lead
   (e) manganese
   (g) oxygen
   (i) tin

   (b) hydrogen
   (d) magnesium
   (f) neon
   (h) sulfur
   (j) zinc

2. *Boiling Point*

    (a)   Place a 400 mL beaker on a wire gauze and support it on a ring stand (Figure 5-2). Add 300 mL of water to the beaker, bring to a boil, and then shut off the burner. Put about 2 mL of methyl alcohol into a test tube and add a boiling chip. Place the test tube in the beaker of water (Figure 5-2). Suspend a thermometer about 1 cm above the liquid. Allow the alcohol to boil in the waterbath for a couple of minutes. Record the temperature ($\pm 0.5^\circ$C) after condensed vapor drops from the tip of the thermometer.

        CAUTION: Methyl alcohol is flammable and the vapors must not come near a flame.

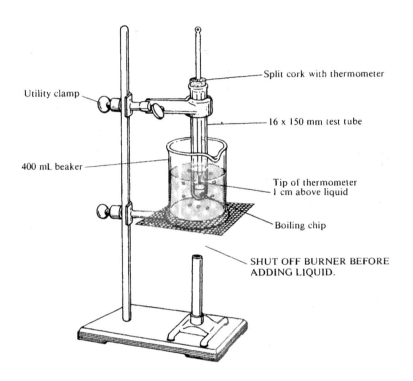

Utility clamp

Split cork with thermometer

16 x 150 mm test tube

400 mL beaker

Tip of thermometer
1 cm above liquid

Boiling chip

SHUT OFF BURNER BEFORE
ADDING LIQUID.

FIGURE 5-2 Apparatus for determining a boiling point.

    (b)   Record the number of an unknown liquid and determine the boiling point as above.

3. *Solubility.* Add 5 mL (1/4 volume) of distilled water into two test tubes. Place a small crystal of iodine in one and a crystal of sucrose (table sugar) in the other and shake the test tubes for a couple of minutes. State whether the substances are *soluble* or *insoluble*.

4. *Miscibility.* Into two test tubes, add 5 mL of distilled water. Add a few drops of methyl alcohol to one and amyl alcohol to the other. Shake briefly and state whether the liquids are *miscible* or *immiscible*.

B. Study of Chemical Properties

1. *Heating Elements*

    (a)   Carefully inspect a piece of copper wire. Heat the wire until it glows red and then allow it to cool. Observe the change and state whether it is *physical* or *chemical*.

(b) Place four small crystals of iodine in a dry beaker, cover with an evaporating dish, and put ice into the dish (Figure 5-3). Support the beaker on a ring stand and heat the iodine slowly until a substance collects on the bottom of the evaporting dish. State whether the change is *physical* or *chemical*.

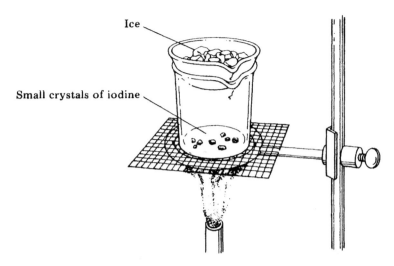

FIGURE 5-3 Apparatus for heating iodine.

2. *Heating Compounds.* Put a pea-sized portion of ammonium bicarbonate into one test tube and potassium bicarbonate into another. Heat each test tube and state whether the change is *physical* or *chemical*.

    NOTE: When heating the test tube, waft to observe any odor.

3. *Solution Reactions*
    (a) Put 2 mL (1/10 volume) of sodium carbonate and sodium sulfate into separate test tubes. Add several drops of dilute hydrochloric acid to each and note any changes. State the change as being *physical* or *chemical*.

         NOTE: No reaction is an example of physical change because the physical properties of mass and volume have increased.

    (b) Put 2 mL (1/10 volume) of sodium nitrate and 2 mL (1/10 volume) lead nitrate into separate test tubes. Add a few drops of potassium iodide and record the change as *physical* or *chemical*.

PRELABORATORY ASSIGNMENT*

1. In your own words define the following terms: chemical changes, chemical properties, compound, element, metal, miscibility, mixture, nonmetal, physical changes, physical properties, sublimation, substance.
2. List several examples of physical properties.
3. List several examples of chemical changes.
4. Estimate the volume of liquid in a test tube that is 1/10 filled. Estimate the volume if the test tube is 1/4 filled with a liquid.
5. What is the purpose of the boiling chip when determining the boiling point of a liquid?
6. What experimental observations suggest a chemical change has taken place?
7. What safety precautions must be observed in this experiment?

*Answers in Appendix II.

DATA TABLE FOR PHYSICAL AND CHEMICAL PROPERTIES

A. Study of Physical Properties

 1. *Observing Physical Properties*

Element	Symbol	Physical State	Color	Metal/Nonmetal
cobalt				
hydrogen				
lead				
magnesium				
manganese				
neon				
oxygen				
sulfur				
tin				
zinc				

 2. *Boiling Point*

  boiling point of methanol (65.0°C)   _____ °C

  boiling point of **UNKNOWN #** _____ _____ °C

 3. *Solubility*

  iodine and water   _____

  sucrose and water   _____

 4. *Miscibility*

  methyl alcohol and water   _____

  amyl alcohol and water   _____

B. Study of Chemical Properties

Procedure	Observation	Change
1. *Heating Elements*		
(a) copper + heat		
(b) iodine + heat		
2. *Heating Compounds*		
ammonium bicarbonate + heat		
potassium bicarbonate + heat		
3. *Solution Reactions*		
(a) sodium carbonate + hydrochloric acid		
sodium sulfate + hydrochloric acid		
(b) sodium nitrate + potassium iodide		
lead nitrate + potassium iodide		

1.   Classify each of the following as an example of a compound, element, or mixture.

   (a)   ethanol, C_2H_5OH _____   (b)   iron, Fe _____

   (c)   glucose, $C_6H_{12}O_6$ _____   (d)   brass, Cu and Zn _____

2.   State whether the following properties are typical of a metal or nonmetal.

   (a)   poor heat conductor _____   (b)   high melting point _____

   (c)   dull/brittle solid _____   (d)   reacts with metals _____

3.   Classify the following properties of sodium metal as physical (ph) or chemical (ch).

   (a)   silver metallic color   _____

   (b)   turns gray in air   _____

   (c)   melting point 98°C   _____

   (d)   density 0.97 g/cm^3 _____

   (e)   reacts explosively with chlorine gas   _____

   (f)   dissolves in water to produce a gas   _____

   (g)   malleable   _____

   (h)   insoluble in ether   _____

   (i)   conductor of electricity   _____

   (j)   forms an oxide when heated with oxygen   _____

4.   Indicate whether these observations are most likely evidence for physical (ph) or chemical (ch) change.

   (a)   Steam condenses to a liquid on a cool surface.   _____

   (b)   Baking soda dissolves in vinegar, generating bubbles.   _____

   (c)   Mothballs gradually disappear at room temperature.   _____

   (d)   Mercury cools to −40°C forming a solid.   _____

   (e)   Steel wool forms small blue-black beads upon heating.   _____

   (f)   Sulfur and iron powders mix together.   _____

   (g)   Iron and sulfur powders change color and texture
         upon heating together.   _____

   (h)   Silver deposits onto a copper wire in a silver nitrate solution.   _____

   (i)   Lead nitrate and hydrochloric acid solutions produce
         an insoluble substance upon mixing.   _____

   (j)   Dry ice, solid carbon dioxide, sublimes.   _____

5. (optional) Using the *Handbook of Chemistry and Physics, The Elements*, find the symbol, boiling point (Bp), density or specific gravity (sp gr), and the year discovered for the following elements.

Element	Symbol	Bp	Density	Year
barium	_____	_____	_____	_____
beryllium	_____	_____	_____	_____
germanium	_____	_____	_____	_____
element 104	_____	N/A	N/A	_____

6. (optional) Using the *Handbook of Chemistry and Physics, Physical Constants of Inorganic Compounds*, find the formula, density or specific gravity, melting point (Mp), and solubility in cold water for the following compounds.

Compound	Formula	Density	Mp	Solubility
lithium hydroxide	_____	_____	_____	_____
mercury(II) fulminate	_____	_____	_____	_____
potassium iodide	_____	_____	_____	_____
sodium chloride	_____	_____	_____	_____

Change of State

OBJECTIVES

1. To gain proficiency in constructing a graph and plotting data points.
2. To determine the freezing point of a compound from the graph of decreasing temperature versus time.
3. To determine the melting points of a known and unknown compound.

DISCUSSION

Matter is defined in terms of having a mass and occupying a volume. The three physical states of matter are solid, liquid, and gas. Matter exists in one of these three physical states depending upon the temperature and atmospheric pressure.

The transition from one physical state to another occurs when the temperature is sufficiently high to supply enough energy so that the individual molecules can overcome the force of attraction to each other. For example, when ice is converted to water through warming, the ice crystal has attained sufficient energy that the molecules of water are no longer held in a rigid network and become free to move past each other. Thus, a liquid is composed of mobile particles that assume the shape of the vessel in which it is contained. Conversely, the opposite process of allowing a liquid to cool results in the loss of heat energy, which leaves the individual molecules with insufficient energy to overcome the force of attraction. Thus, the molecules become fixed in position and a solid is formed.

At that temperature where a liquid changes to a solid, two physical states are present simultaneously. This temperature is referred to as the *freezing point*; or, if a solid is changing into a liquid it is called the *melting point*. The freezing point and melting point have the same value, theoretically.

In this experiment, liquid paradichlorobenzene will be allowed to cool to a solid while the temperature/time relationship is recorded. The data will be plotted and the graph labeled as a cooling curve. The temperature should remain constant as the liquid solidifies. This is because heat is released as the mobile liquid molecules stop moving to form a rigid solid crystal. Figure 6-1 shows a typical cooling curve.

The compound is initially heated to a liquid and then allowed to cool. As the compound cools, crystals begin to form. After a few minutes the crystals become a solid mass. The liquid has undergone a change of state to a solid. The temperature/time data is carefully recorded and the

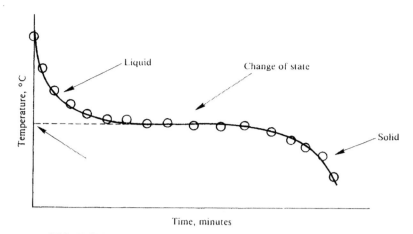

FIGURE 6-1   Cooling curve demonstrating change of state.

data points plotted. The plateau in the cooling curve is extrapolated to the vertical axis. The freezing point (Fp) corresponds to the temperature of the plateau.

The apparatus for determining the cooling curve is shown in Figure 6-2.

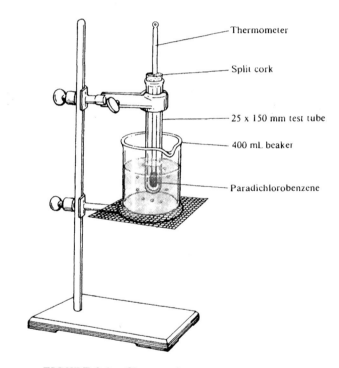

FIGURE 6-2   Change of state apparatus.

In the second procedure of this experiment a melting point is determined. A small sample of compound is rapidly heated until it is observed to liquefy. The temperature range over which the compound melts is recorded; for example, 65-75°C. A second trial is repeated for greater accuracy. The waterbath is heated rapidly to 60°C and then slowly until the compound melts. This second trail should produce an accurate melting point with a 1-2°C range; for example 69-71°C.

EQUIPMENT AND CHEMICALS

- wire gauze
- utility clamp
- 110°C thermometer with split cork
- 400 mL beaker
- 25 × 150 mm test tube containing 20 g of paradichlorobenzene

- mortar and pestle
- capillary tubes
- rubber bands
- biphenyl (diphenyl)
- melting point unknowns

PROCEDURE

A. Cooling Curve—A Graphing Exercise The instructor may wish to have students work in partners. One student should set up the apparatus and record data while the partner heats the paradichlorobenzene and later takes temperature readings.

1. Set up the apparatus as shown in Figure 6-2. Add 300 mL of distilled water to the 400 mL beaker. Heat the water to 40°C and shut off the burner.

2. Obtain a test tube containing paradichlorobenzene and melt the compound by immersing the test tube in a beaker of boiling water. Insert a thermometer into the melted compound and continue heating until the temperature is well above 65°C.

   NOTE: For this purpose, it is convenient to prepare a large waterbath heated on a hotplate. Allow the water to boil gently.

3. Rapidly transfer the test tube with thermometer into the 400 mL beaker of heated water. Support the test tube with a utility clamp and hold the thermometer with a split cork as shown in Figure 6-2.

4. Begin thermometer readings when the temperature drops to 65.0°C. Record temperature every 30 seconds. Continue recording temperature/time data until the compound is solidified.

   NOTE: To ensure a correct temperature reading (±0.5°C), it is advisable to stir the compound as it cools. Before taking the temperature, move the thermometer up and down to mix the compound.

5. From the recorded data, graph a cooling curve by plotting temperature on the vertical axis and time on the horizontal axis. Circle each point and construct a smooth curve. Extrapolate the temperature plateau to the vertical axis to find the freezing point of the compound. Show the graph to the instructor who may request a second trial.

   NOTE: At the conclusion of this part of the experiment, the thermometer will be frozen in the paradichlorobenzene. Do not attempt to pull it out because the thermometer can break. Simply melt the compound and the thermometer is released naturally. Also, do not pour out the liquid paradichlorobenzene.

B. Melting Points

1. Seal one end of capillary tube with a burner flame. Dab the open end into a small sample of biphenyl. Invert the capillary and lightly tap the sealed end to pack the sample. Repeat this process until a 5 mm sample is packed at the sealed end of the capillary.

   NOTE: If the crystals of the compound are large, grind them smaller using the mortar and pestle. Packing the crystals may be done by vertically dropping the sealed end of the capillary through a half meter of 6 mm glass tubing onto the lab bench top.

2. Set up an apparatus as shown in Figure 6-3. Add 300 mL of distilled water to the 400 mL beaker. Attach the capillary at the end of the thermometer with a rubber band and place in the beaker.

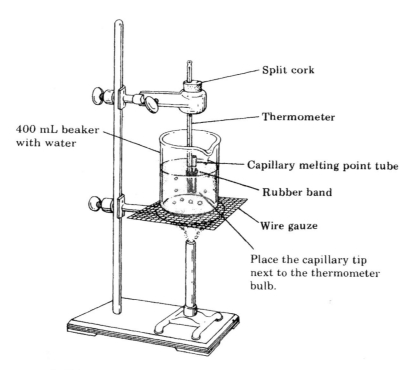

— Split cork

— Thermometer

400 mL beaker with water

— Capillary melting point tube

— Rubber band

— Wire gauze

Place the capillary tip next to the thermometer bulb.

FIGURE 6-3 Apparatus for determining a melting point.

3. Rapidly heat the water in the beaker until the biphenyl melts. Observe the approximate melting point and record the range of temperature in the Data Table.
4. Prepare another capillary tube and heat rapidly until the temperature is within 10°C of the melting point. Then slowly continue to heat in order to determine the melting point accurately. Record the melting point range from the first sign of melting until the compound has completely melted. The reference value is given in the Data Table for comparison.
5. Obtain an unknown compound and record the unknown number. Determine the melting point for the unknown as above.

PRELABORATORY ASSIGNMENT*

1. In your own words define the following terms: abscissa, change of state, freezing point, ordinate axis, origin, physical state.
2. Why should distilled water be used in the waterbath?
3. The freezing point of paradichlorobenzene corresponds to which point on the cooling curve?
4. After gathering data for the cooling curve, how is the thermometer removed from the solid paradichlorobenzene?
5. In determining a melting point of a compound, why are two trials performed?
6. While performing the second melting point trial the compound begins to liquefy at 68.0°C and liquefies completely at 69.5°C. Report the Mp.
7. A compound contained in a capillary tube is placed in the waterbath and liquefies even before heating. Name two possible sources of the problem.
8. What safety precautions must be observed in this experiment?

*Answers in Appendix II.

DATA TABLE FOR CHANGE OF STATE

A. Cooling Curve—A Graphing Exercise

Temperature	Time	Observation
65.0°C	0:00	liquid
	0:30	
	1:00	
	1:30	
	2:00	
	2:30	
	3:00	
	3:30	
	4:00	
	4:30	
	5:00	
	5:30	
	6:00	
	6:30	
	7:00	
	7:30	
	8:00	
	8:30	
	9:00	
	9:30	
	10:00	

B. Melting Points

	Rapid Trial	Trial 2
Mp of biphenyl (69-71°C)	_____ °C	_____ °C
Mp of UNKNOWN # _____	_____ °C	_____ °C

Cooling Curve   Trial 2

Temperature	Time	Observation
65.0°C	0:00	liquid
	0:30	
	1:00	
	1:30	
	2:00	
	2:30	
	3:00	
	3:30	
	4:00	
	4:30	
	5:00	
	5:30	
	6:00	
	6:30	
	7:00	
	7:30	
	8:00	
	8:30	
	9:00	
	9:30	
	10:00	

A. COOLING CURVE Trial 1   Freezing Point: _____ °C

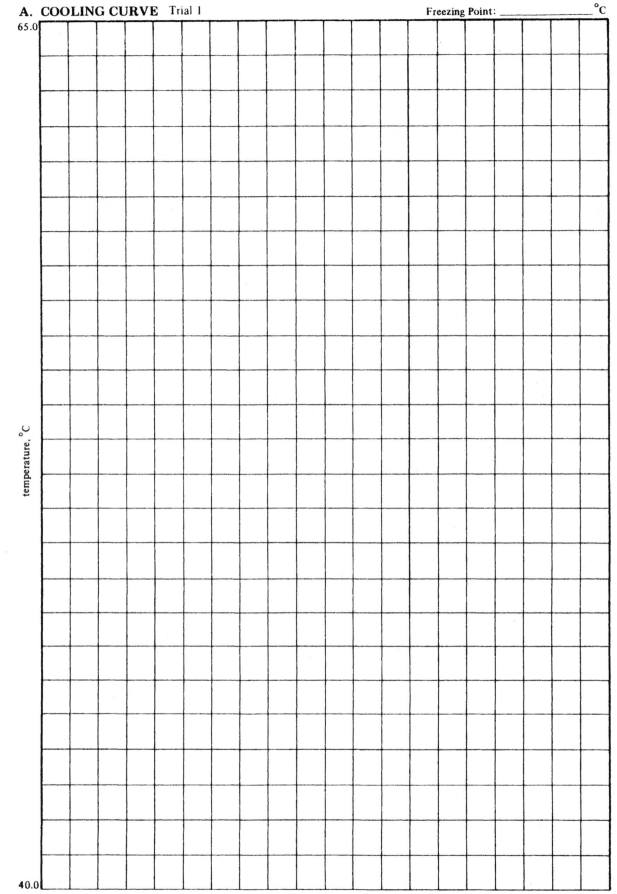

temperature, °C

65.0

40.0

0   time, minutes   10

B. COOLING CURVE Trial 2   Freezing Point: _____ °C

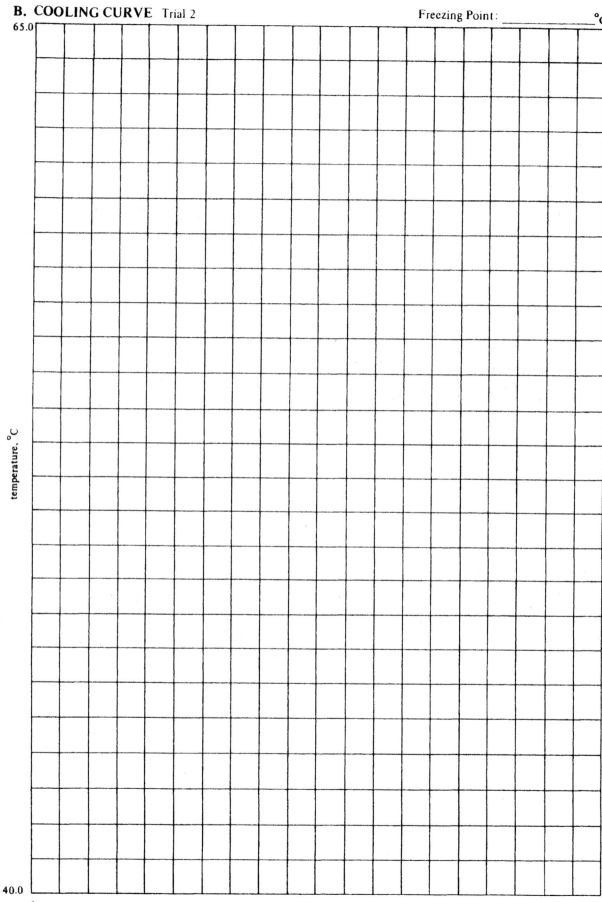

65.0

temperature, °C

40.0

0   time, minutes   10

1.   Methanol, or wood alcohol, solidifies at −94°C. What is the freezing point on the Kelvin and Fahrenheit scales?

_____

_____

2.   Urea, excreted in the urine of animals, liquefies at 135°C. What is the melting point on the Kelvin and Fahrenheit scales?

_____

_____

3.   The following graph depicts a heating curve for acetic acid:

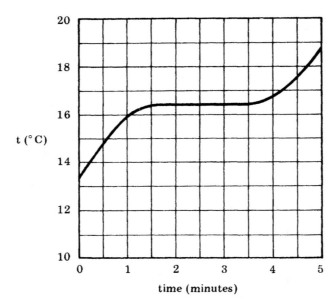

From the graph, estimate the melting point to the nearest 0.1°C for acetic acid.

_____

4. Using the following data for the compound benzene, graph the cooling curve.

Temperature (°C)	Time (minutes)
15.0	0:00
13.4	0:30
10.2	1:00
7.8	1:30
6.6	2:00
5.8	2:30
5.6	3:00
5.5	3:30
5.5	4:00
5.5	4:30
5.4	5:00

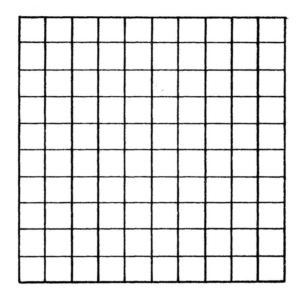

From the graph, estimate the freezing point to the nearest 0.1°C for benzene.

_____

5. (optional) Using the *Handbook of Chemistry and Physics, Physical Constants of Organic Compounds*, find the melting points of

(a) acetic acid   (b) benzene

_____

Periodic Classification of the Elements

OBJECTIVES

1. To study the similarity of properties for groups of elements in the Periodic Table.
2. To observe the flame tests and solution reactions of some alkali and alkaline earth elements.
3. To become familiar with the reactions of chlorine water and the halides.
4. To analyze an unknown solution containing an alkali or alkaline earth element and a halide.

DISCUSSION

It was not until the beginning of the nineteenth century that the clear distinction between elements and compounds was made. At that time, the study of the composition of substance was flourishing. Chemists were discovering a vast body of knowledge about the physical and chemical properties of elements and compounds. As a result of this rapid accumulation of data, several systematic methods of classification for the elements were attempted. In 1869, the Russian chemist Dmitri Mendeleev suggested the systematic arrangement of elements according to increasing atomic mass. Mendeleev had the brilliant foresight to predict yet undiscovered elements and he represented these elements by spaces in his proposed Periodic Table of elements. The construction of the table and the ordering of elements into groups and periods was based upon observed physical and chemical properties. Mendeleev was not only able to predict the existence of six unknown elements, but also many properties of these elements.

In 1913 Harry Moseley, a 25-year-old English physicist, was working in the laboratory of Ernest Rutherford at the University of Manchester. Moseley examined X-ray emission spectra and concluded that the elements should actually be arranged in order of increasing atomic number rather than atomic mass. With few exceptions, an increase in atomic number is paralleled by an increase in atomic mass. However, the concept of increasing atomic number much more clearly explains periodic properties. The periodic recurrence of similar physical and chemical properties, based upon atomic number, is referred to as the *Periodic Law*.

The elements in the Periodic Table (Figure 7-1) are arranged in a sequence of columns and rows. The elements in vertical columns are called groups or families and possess similar chemical properties. The horizontal rows are termed periods or series. There is a general trend in physical properties for elements within a group. For example, density usually increases from the top to bottom of a group. Elements that belong to the same family also give similar chemical reactions.

GROUPS

PERIODS	1 IA	2 IIA	3 IIIB	4 IVB	5 VB	6 VIB	7 VIIB	8	9 VIII	10	11 IB	12 IIB	13 IIIA	14 IVA	15 VA	16 VIA	17 VIIA	18 VIIIA
1	H 1																	He 2
2	Li 3	Be 4											B 5	C 6	N 7	O 8	F 9	Ne 10
3	Na 11	Mg 12											Al 13	Si 14	P 15	S 16	Cl 17	Ar 18
4	K 19	Ca 20	Sc 21	Ti 22	V 23	Cr 24	Mn 25	Fe 26	Co 27	Ni 28	Cu 29	Zn 30	Ga 31	Ge 32	As 33	Se 34	Br 35	Kr 36
5	Rb 37	Sr 38	Y 39	Zr 40	Nb 41	Mo 42	Tc 43	Ru 44	Rh 45	Pd 46	Ag 47	Cd 48	In 49	Sn 50	Sb 51	Te 52	I 53	Xe 54
6	Cs 55	Ba 56	*La 57	Hf 72	Ta 73	W 74	Re 75	Os 76	Ir 77	Pt 78	Au 79	Hg 80	Tl 81	Pb 82	Bi 83	Po 84	At 85	Rn 86
7	Fr 87	Ra 88	**Ac 89	Rf 104	Ha 105	Unh 106	Uns 107	Uno 108	Une 109									

TRANSITION ELEMENTS

*Lanthanide series

Ce 58	Pr 59	Nd 60	Pm 61	Sm 62	Eu 63	Gd 64	Tb 65	Dy 66	Ho 67	Er 68	Tm 69	Yb 70	Lu 71
140.12		144.24		150.4		157.25			164.930		168.934	173.04	174.967

**Actinide series

Th 90	Pa 91	U 92	Np 93	Pu 94	Am 95	Cm 96	Bk 97	Cf 98	Es 99	Fm 100	Md 101	No 102	Lr 103
232.038		238.029		244	243			251	254	(257)	(256)	(255)	(257)

FIGURE 7-1   A modern Periodic Table arranged into group of elements having similar properties.

In this experiment we will observe flame tests and solution reactions of selected alkali and alkaline earth elements. A *flame test* is a diagnostic test that is performed by placing a small amount of solution on the coiled tip of a wire and holding the wire over a hot flame to observe the color produced (Figure 7-2). For example, sodium solutions give a yellow flame test, copper solutions a green flame test, and the flame test for silver solutions is not visible.

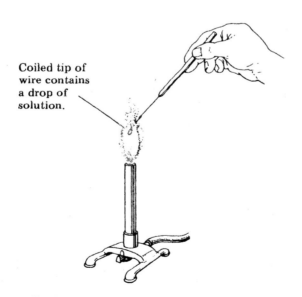

Coiled tip of wire contains a drop of solution.

FIGURE 7-2   The technique of flame testing.

Although flame tests are usually specific for each element, they can be misleading. Sodium is always present as an impurity. Therefore, a flame test will always give a yellow flame. However, the intensity of the yellow flame is less for an impurity than for a sodium compound and the distinction can be made with a little practice. Oftentimes the color of the flame test is similar for two different elements. In these cases, an element is confirmed by directly comparing the flame test of an unknown to a known solution.

EQUIPMENT AND CHEMICALS

Equipment

- 16 × 150 mm test tubes (6)
- test tube rack
- test tube brush
- flame test wire (chromel, nichrome, or platinum)
- wash bottle

Chemicals

- ammonium carbonate solution, 0.5 M $(NH_4)_2 CO_3$
- ammonium phosphate solution, 0.5 M $(NH_4)_2 HPO_4$
- ammonium sulfate solution, 0.5 M $(NH_4)_2 SO_4$
- Barium solution, 0.5 M $BaCl_2$

- Calcium solution, 0.5 M $CaCl_2$
- Lithium solution, 0.5 M LiCl
- Potassium solution, 0.5 M KCl
- Sodium solution, 0.5 M NaCl
- Strontium solution, 0.5 M $SrCl_2$
- Chloride solution, 0.5 M NaCl
- Bromide solution, 0.5 M NaBr
- Iodide solution, 0.5 M NaI
- hexane, C_6H_6
- dilute nitric acid, 6 M HNO_3
- chlorine water (bleach)
- unknown solutions containing one of the above alkali or alkaline earth elements and a halide, 0.5 M concentration

PROCEDURE

A. Flame Tests for the Alkali and Alkaline Earth Elements

1. Place six test tubes in a test tube rack. Add 2 mL (1/10 test tube) of the following solutions into separate test tubes: barium, calcium, lithium, potassium, sodium, strontium (see Figure 7-3).

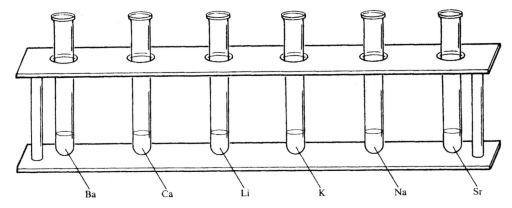

FIGURE 7-3   Test tube rack with solutions of alkali and alkaline earth elements.

2. Obtain a flame test wire and make a small loop in the end. Heat the loop with a burner at the tip of the blue cone. Continue to heat the wire until there is no longer any color produced in the flame. To avoid contamination, do not touch the clean wire.
3. Dip the clean wire into the test tube containing the barium solution. Place the loop at the tip of the flame. Record your observation. Clean the wire and repeat the flame test for the solutions containing calcium, lithium, potassium, sodium, and strontium.

To clean a used wire, dip the wire in concentrated hydrochloric acid and heat the wire to red heat. In some instances, it may be necessary to repeat this operation.

B. Reactions of the Alkali and Alkaline Earth Elements

1. Add 1 mL of ammonium carbonate solution to each test tube. If a precipitate forms, record *ppt* in the Data Table. If there is no reaction, record *NR*.
2. Clean the test tubes and rinse with distilled water. Put 2 mL (1/10 test tube) of the barium, calcium, lithium, potassium, sodium, strontium solutions into separate test tubes. Add 1 mL of ammonium phosphate solution into each test tube. Record your observations in the Data Table.
3. Clean the test tubes and put 2 mL of the barium, calcium, lithium, potassium, sodium, and strontium solutions into separate test tubes. Add 1 mL of ammonium sulfate solution into each test tube and record your observations.

C. Reactions of the Halides

1. Place three test tubes in a test tube rack. Add 2 mL (1/10 test tube) of the following solutions into separate test tubes: chloride, bromide, iodide.
2. Into each test tube add 2 mL of hexane, 2 mL of chlorine water, and a drop of nitric acid.
3. Shake each test tube and observe the color of the upper hexane layer.

   NOTE: Dispose of these solutions in a special waste container for organic chemicals.

D. Analysis of an Unknown Solution

1. Record the unknown number of a solution assigned by the instructor. Perform a flame test on the unknown solution and record your observation in the Data Table.
2. Put 2 mL of unknown solution into each of three test tubes. Add 1 mL of ammonium carbonate to the first; 1 mL of ammonium phosphate to the second; and 1 mL of ammonium sulfate to the third. Record your observations in the Data Table.
3. Put 2 mL of unknown solution into a test tube. Add 2 mL of hexane, 2 mL of chlorine water, and a drop of nitric acid. Shake the test tube and record the color of the upper hexane layer in the Data Table.
4. Compare the flame test and solution reactions of the unknown solution to the six known solutions (Procedures A and B). Deduce which of the alkali or alkaline earth elements is present in the unknown solution.
5. Compare the halide test of the unknown solution to the three known solutions (Procedure C). Deduce which of the halides is present in the unknown solution.

PRELABORATORY ASSIGNMENT*

1. In your own words define the following terms: alkali metals, alkaline earth metals, flame test, halide, immiscible, and precipitate.
2. In this experiment which three alkali elements are investigated? Which three alkaline earth elements? Which three halides?
3. What difficulties arise in interpreting a flame test?
4. Are water and hexane miscible? Which is the hexane layer? In which layer do you observe the halide test?
5. What safety precautions should be observed in this experiment?

*Answers in Appendix II.

Structure of Compounds

OBJECTIVES

1. To construct models of various compounds.
2. To be able to sketch each model showing its three-dimensional structure.
3. To be able to draw the structural formula for each compound based upon the model.
4. To write the electron-dot formula for each structural formula.
5. To verify the electron-dot formula by a valence electron check.
6. To write the structural and electron-dot formulas for models of unknown compounds.

DISCUSSION

A chemical bond is the force of attraction that holds a compound together. In a covalent bond two nonmetallic atoms are held together by the sharing of valence electrons. The valence electrons are the electrons furthest from the nucleus and occupy the highest s and p sublevels. The number of valence electrons can be determined by writing out the electron configuration of the element. More simply, the valence electrons are found by inspecting the Periodic Table and noting the group number of the element.

PROBLEM EXAMPLE 8-1

Find the number of valence electrons for an atom of each of the following elements: (a) H; (b) C; (c) N; (d) O; (e) Cl, Br, I.

Solution: (a) The element hydrogen is in Group IA. Therefore, hydrogen has one valence electron.

(b) Carbon is in Group IVA; it has four valence electrons.

(c) Oxygen is in Group VIA; it has six valence electrons.

(d) Nitrogen is in Group VA and has five valence electrons. However, under ordinary conditions only three of nitrogen's valence electrons are shared in chemical bonds. The remaining two do not usually bond and are termed a nonbonding electron pair.

(e) Chlorine, bromine, and iodine are in Group VIIA and thus each has seven valence electrons.

In summary, the group number of a representative element corresponds to the number of valence electrons. This does not apply to the transition elements which are filling the *d* sublevel with electrons.

The central objective of this experiment is to deduce the electron-dot formula of a compound from a model. The model is constructed from balls and sticks. Each ball represents an atom and each stick a single covalent bond. A single bond shares two electrons so each peg represents an electron pair.

A double bond shares two pairs of electrons. A model compound is constructed using two springs to represent the double bond. The springs replace the peg and are flexible enough to connect two holes in each ball. Each spring represents an electron pair.

A triple covalent bond contains three pairs of electrons. A model compound having a triple bond uses three springs to join two balls together. Each spring represents one pair of electrons.

The following examples illustrate model compounds having single, double, and triple bonds.

PROBLEM EXAMPLE 8-2

The model of a water molecule is sketched below. Draw (a) the structural formula and (b) the electron-dot formula corresponding to the model and (c) verify the electron-dot formula by checking the total number of electron dots against the sum of all valence electrons.

Water, H₂O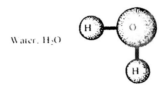

Solution: (a) Each stick represents a single bond; therefore the structural formula is

$$H - O$$
$$|$$
$$H$$

(b) A dash in the structural formula indicates an electron pair, thus

$$H : O$$
$$..$$
$$H$$

Each hydrogen shares two electrons which is its maximum. However, the oxygen requires an octet of electrons and shares only four in the above diagram. Therefore, we will add two more pairs of electrons to complete the octet. The electron-dot formula is

$$\overset{..}{H : O :}$$
$$..$$
$$H$$

(c) To verify the above formula we will add up the valence electrons from each atom in the molecule. Recall that hydrogen is in Group IA and oxygen is in Group VIA.

$$2\,H\,(2 \times 1e) = 2e^-$$
$$1\,O\,(1 \times 6e) = 6e^-$$
$$sum\ of\ valence\ electrons = 8e^-$$

There are eight dots used to write the electron-dot formula. Since this equals the number of valence electrons, the electron-dot formula is correct.

PROBLEM EXAMPLE 8-3

The three-dimensional model of chloroform is sketched below. Draw (a) the structural formula and (b) the electron-dot formula. Each atom (excluding H) should be surrounded by an octet of electrons. (c) Verify the electron-dot formula by checking the total number of electron dots against the sum of all valence electrons.

Chloroform, $CHCl_3$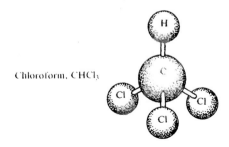

Solution: (a) Each stick represents a single bond so the structural formula is

$$
\begin{array}{c}
\text{H} \\
| \\
\text{Cl} - \text{C} - \text{Cl} \\
| \\
\text{Cl}
\end{array}
$$

(b) Each dash in the structural formula indicates an electron pair; therefore,

$$
\begin{array}{c}
\text{H} \\
\cdot\cdot \\
\text{Cl} : \text{C} : \text{Cl} \\
\cdot\cdot \\
\text{Cl}
\end{array}
$$

Hydrogen and carbon are complete as shown; two electrons and eight electrons respectively. However, each chlorine also requires an octet which we will complete as follows:

$$
\begin{array}{c}
\text{H} \\
\cdot\cdot \quad \cdot\cdot \quad \cdot\cdot \\
:\text{Cl} : \text{C} : \text{Cl}: \\
\cdot\cdot \quad \cdot\cdot \quad \cdot\cdot \\
:\text{Cl}: \\
\cdot\cdot
\end{array}
$$

(c) To verify the above electron-dot formula we will find the sum of all valence electrons.

$$
\begin{array}{rl}
1 \text{ H } (1 \times 1e) = & 1e^- \\
1 \text{ C } (1 \times 4e) = & 4e^- \\
3 \text{ Cl } (3 \times 7e) = & 21e^- \\
\hline
\textit{sum of valence electrons} = & 26e^-
\end{array}
$$

There are 26 valence electrons and 26 dots used in the electron-dot formula. Thus, the formula is verified.

PROBLEM EXAMPLE 8-4

A molecular model of formaldehyde is sketched below. Draw the (a) structural formula and (b) electron-dot formula. (c) Find the sum of all valence electrons to verify the electron-dot formula.

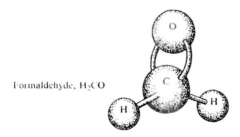

Formaldehyde, H₂CO

Solution: (a) Two springs joining the carbon and oxygen atoms represent a double bond. The structural formula is

$$\begin{array}{c} O \\ \parallel \\ H-C-H \end{array}$$

(b) Each single bond contains one electron pair and the double bond two electron pairs.

$$\begin{array}{c} O \\ :: \\ H:C:H \end{array}$$

Hydrogen shares two electrons and is complete. Carbon shares a total of eight electrons and satisfies the octet rule. Oxygen has only four of the eight electrons necessary to complete the octet. Therefore, we will add two unshared electron pairs.

$$\begin{array}{c} \cdot\cdot \\ O: \\ :: \\ H:C:H \end{array}$$

(c) We verify the above electron-dot formula as follows:

$$2\,H\,(2 \times 1e) = 2e^-$$
$$1\,C\,(1 \times 4e) = 4e^-$$
$$1\,O\,(1 \times 6e) = \underline{6e^-}$$
$$\textit{sum of valence electrons} = 12e^-$$

The 12 valence electrons equal the 12 electron dots so the formula is verified.

PROBLEM EXAMPLE 8-5

A model of hydrogen cyanide is sketched below. Write (a) the structural formula and (b) the electron-dot formula. (c) Verify the electron-dot formula.

Hydrogen cyanide, HCN

Solution: (a) The three springs connecting the carbon and nitrogen represent a triple pair of electrons.

$$H - C \equiv N$$

(b) We can write an electron-dot formula after realizing the triple bond contains three electron pairs.

$$H : C : : : N$$

In the above formula nitrogen shares only six electrons. Therefore, we must add one unshared electron pair.

$$H : C : : : N :$$

(c) Let's verify the above electron-dot formula.

$$
\begin{aligned}
1\ H\ (1 \times 1e) &= 1e^- \\
1\ C\ (1 \times 4e) &= 4e^- \\
1\ N\ (1 \times 5e) &= \underline{5e^-} \\
\textit{sum of valence electrons} &= 10e^-
\end{aligned}
$$

The 10 valence electrons correspond to the $10e^-$ dots; thus the electron-dot formula is verified.

EQUIPMENT AND CHEMICALS

- Molecular model kits

Directions for Using Molecular Models When constructing a model, a hole in a ball represents a missing electron that is necessary in order to complete an octet. If two balls are joined together with a single connector, the connector represents a bond composed of two electrons. If two balls are joined together by two connectors, a double bond is indicated and represents four bonding electrons. Three connectors joining two balls represents a triple bond and a total of six electrons. The six electrons are perhaps more precisely referred to as three pairs of bonding electrons.

one connector—single bond (electron pair)
two connectors—double bond (two electron pairs)
three connectors—triple bond (three electron pairs)

As the model is constructed for a compound, all the holes in each ball should be filled with a connector. (Nitrogen may be an exception.) The color code for each ball is as follows:

white or yellow ball—hydrogen (one hole)
black ball—carbon (four holes)
red ball—oxygen (two holes)
blue ball—nitrogen (four or five holes)
green ball—chlorine (one hole)
orange ball—bromine (one hole)
purple ball—iodine (one hole)

PROCEDURE

DIRECTIONS FOR COMPLETING THE DATA TABLE:

1. Construct molecular models for the compounds below. Sketch the model in the Data Table consistent with its three-dimensional structure.
2. Draw the structural formula for each compound.
3. Write the electron-dot formula corresponding to the structural formula. Each atom should be surrounded by an octet of electrons.

   NOTE: Hydrogen is an exception as it has one energy sublevel and shares only two electrons.

4. Verify each electron-dot formula by summing the valence electrons; then count the number of electron dots.

A. Compounds with Single Bonds

(a) H_2	(b) Cl_2
(c) Br_2	(d) I_2
(e) HCl	(f) HBr
(g) IBr	(h) ICl
(i) CH_4	(j) CH_2Cl_2
(k) $HOCl$	(l) H_2O_2
(m) NH_3	(n) N_2H_4
(o) NH_2OH	(p) CH_3NH_2

B. Compounds with a Double Bond

(a) O_2	(b) C_2H_4
(c) $HONO$	(d) $HCOOH$
(e) C_2HCl_3	

C. Compounds with a Triple Bond

(a) N_2	(b) C_2H_2
(c) $HOCN$	

D. Compounds with Two Double Bonds

(a) CO_2	(b) C_3H_4
(c) C_2H_2O	

E. **Unknown Compounds** The instructor will provide molecular models of unknown compounds. Draw the structural formula of each unknown compound. Write the electron-dot formula corresponding to each structural formula.

PRELABORATORY ASSIGNMENT*

1. In your own words define the following terms: double bond, electron-dot formula, octet rule, single bond, structural formula, triple bond, unshared pair of electrons, valence electrons.

2. Using the Periodic Table predict the number of valence electrons for the following elements: H, C, N, O, Cl.

3. What do each of the following represent in the model kit?
   (a) yellow ball
   (b) black ball
   (c) red ball
   (d) one connector
   (e) two connectors
   (f) three connectors

4. Draw the structural formula and electron-dot formula for each of the following model sketches.

   (a) IBr

   (b) CH₃Cl

   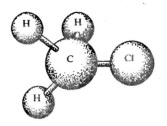

   (c) Cl₂CO

   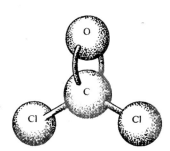

5. Perform a valence electron check on each of the examples in the preceding questions.

*Answers in Appendix II.

NAME ____________________

C. Compounds with Triple Bonds

(a) N_2

(b) C_2H_2

(c) HOCN

D. Compounds with Two Double Bonds

(a) CO_2

(b) C_3H_4

(c) C_2H_2O

Compound	Model	Structural Formula	Electron-dot Formula	Valence Electrons Total

E. Unknown Compounds

#1 ☐

#2 ☐

#3 ☐

#4 ☐

Cation Analysis

OBJECTIVES

1. To observe the chemical behavior of barium, calcium, and magnesium ions.
2. To analyze an unknown solution for one or more of the following cations: Ba^{2+}, Ca^{2+}, Mg^{2+}.
3. To develop the following laboratory skills: centrifuging, flame testing, and using litmus paper.

DISCUSSION

The analysis of a solution for the cations present is based upon the separation and identification of each ion. Separation is achieved by selecting a reagent that reacts with one ion in solution but not the others. For example, barium ions can be precipitated with ammonium sulfate while calcium and magnesium ions do not react.

When a barium cation and a sulfate anion are together in the same solution, a precipitate will form because barium sulfate is insoluble. If calcium or magnesium ions are in the solution, no precipitate will form because calcium sulfate and magnesium sulfate are soluble. Figure 9-1 illustrates this separation of cations.

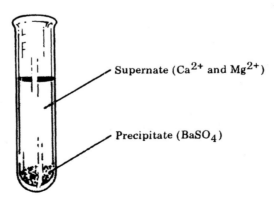

Supernate (Ca^{2+} and Mg^{2+})

Precipitate ($BaSO_4$)

FIGURE 9-1   The test tube after precipitating Ba^{2+} and centrifuging the known solution.

If you have a solution containing several cations, it is usually possible to select a reagent that will form an insoluble product with one of the cations but not with the others. By centrifuging the solid particles in solution, you can separate the precipitated cations from the remaining ions in the solution. The supernate is then poured off and the analysis continues for the remaining cations in the solution.

In this experiment you will separate and identify Ba^{2+}, Ca^{2+}, and Mg^{2+}. First, a known solution containing all three cations will be analyzed to develop techniques and observe reactions. Second, an unknown solution with one or more of the three cations will be analyzed to determine the ions present.

A *semimicro* approach is employed in this experiment. That is, small test tubes and drop amounts of reagents will be used. Precipitates will be separated by centrifuging. Centrifuging forces the solid particles to the bottom of the test tube where they pack firmly. This allows the supernate to be poured off free of particles.

Flame testing is a technique you will use to confirm the presence of an ion. A *flame test* is performed by dipping a wire into a solution and then holding the wire in a hot flame while observing the color produced (Figure 9-2). Many elements produce colored flames. For example, sodium is yellow, potassium is violet, and copper is green. Since sodium is an ever-present impurity, flame tests invariably are contaminated by the yellow sodium flame.

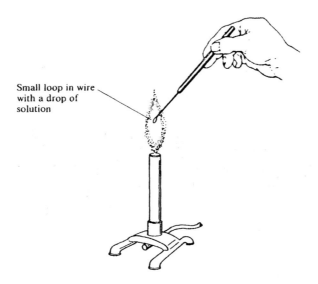

Small loop in wire with a drop of solution

FIGURE 9-2 Flame testing produces a characteristic color for many elements.

Litmus paper is used to determine whether a solution is acidic or basic. A stirring rod is placed into the solution and touched to the litmus paper. Acidic solutions turn blue litmus paper red. Basic solutions turn red litmus paper blue (Figure 9-3).

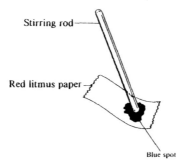

Stirring rod

Red litmus paper

Blue spot

FIGURE 9-3 Testing a base using red litmus paper. A blue spot is produced.

We will begin the analysis with a known solution containing Ba^{2+}, Ca^{2+}, and Mg^{2+}. Ammonium sulfate is added and the separation begins. Figure 9-4 presents an overview of the analysis. In Step 1 Ba^{2+} is confirmed. Step 2 confirms Ca^{2+} and Step 3 confirms Mg^{2+}.

EQUIPMENT AND CHEMICALS

Equipment

- 13 X 100 mm test tubes (3)
- thin glass stirring rod
- wash bottle with distilled water
- centrifuge
- flame test wire (chromel, nichrome, or platinum)
- red litmus paper

Chemicals

- known solution containing Ba^{2+}, Ca^{2+}, and Mg^{2+} (0.1 M $BaCl_2$, $CaCl_2$, $MgCl_2$)

- ammonium sulfate solution, 0.2 M $(NH_4)_2SO_4$
- ammonium oxalate solution, 0.2 M $(NH_4)_2C_2O_4$
- sodium monohydrogen phosphate solution, 0.2 M Na_2HPO_4
- dilute hydrochloric acid, 6 M HCl
- dilute sodium hydroxide, 6 M NaOH
- magnesium indicator (0.1 g p-nitrobenzeneazoresorcinol in 1 L of 0.025 M NaOH)
- unknown solutions (containing one or more of the above cations in 0.1 M concentration)

PROCEDURE

General Directions   Clean three test tubes and a stirring rod with distilled water. Label the test tubes #1, #2, and #3. As a solution is analyzed, record the color of each precipitate in the Data Table.

A. Analysis of a Known Solution

1. *Separation and Identification of Ba^{2+} in a Known Solution of Ba^{2+}, Ca^{2+}, and Mg^{2+}*
    (a) Place 10 drops of the known solution into test tube #1. Add 10 drops of ammonium sulfate, $(NH_4)_2SO_4$, and mix with a stir rod.   *Turn white PPt*

    NOTE: A white precipitate, $BaSO_4$, suggests Ba^{2+} is present.

    (b) Centrifuge and then test for completeness of precipitation by adding another drop of ammonium sulfate. Pour off the supernate into test tube #2 and save for Step 2.
    (c) Add 5 drops of dilute hydrochloric acid, HCl, into test tube #1 and stir thoroughly. Clean a flame test wire with hydrochloric acid and dip it into the solution. Place the wire loop in a hot flame and record the color.   *No flame change*

    NOTE: A green flame confirms Ba^{2+} is present.

2. *Separation and Identification of Ca^{2+} in a Known Solution of Ca^{2+} and Mg^{2+}*
    (a) Add 10 drops of ammonium oxalate, $(NH_4)_2C_2O_4$, to the solution in test tube #2.   *Yes ppt*

    NOTE: A white precipitate, CaC_2O_4, suggests Ca^{2+} is present.

    (b) Centrifuge and then test for completeness of precipitation by adding another drop of ammonium oxalate. Pour off the supernate into test tube #3 and save for Step 3.

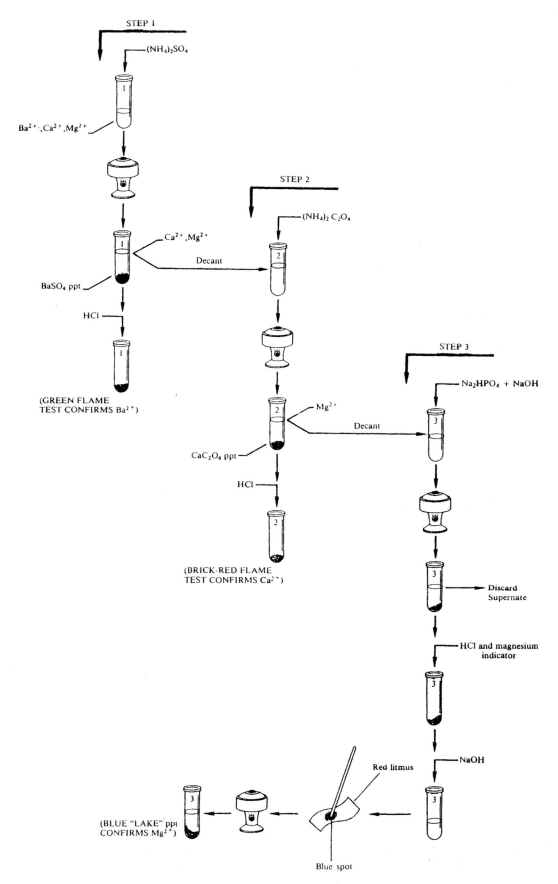

FIGURE 9-4 Separation and identification of the barium, calcium, and magnesium cations.

(c) Add 5 drops of dilute hydrochloric acid into test tube #2 and stir thoroughly. Clean a flame test wire with dilute HCl and dip the wire into the solution. Place the wire loop in a hot flame and record the color. *no change in flame change*

NOTE: A brick-red flame confirms Ca^{2+} is present.

3. *Identification of Mg^{2+} in a Known Solution*
   (a) Add 10 drops of sodium monohydrogen phosphate, Na_2HPO_4, to the solution in test tube #3. Add sodium hydroxide, NaOH, and stir.

   NOTE: A white precipitate suggests Mg^{2+} is present. *– no ppt*

   (b) Centrifuge and discard the supernate.
   (c) Dissolve the precipitate with dilute hydrochloric acid in test tube #3. Add a few drops of magnesium indicator. Add sodium hydroxide, NaOH, dropwise until the solution tests basic on red litmus paper. Centrifuge.

   NOTE: A blue "lake" precipitate confirms Mg^{2+} is present. *– Turns blue confirms mg*

B. Analysis of an Unknown Solution   *sol #47*

1. *Separation and Identification of Ba^{2+} in an Unknown Solution*
   (a) Place 10 drops of unknown solution into test tube #1. Add 10 drops of ammonium sulfate, $(NH_4)_2SO_4$, and mix with a stir rod.

   NOTE: If there is no precipitate, Ba^{2+} is absent. Go directly to Step 2. *↳ yes – ppt*

   (b) Centrifuge and then test for completeness of precipitation by adding another drop of ammonium sulfate. Pour off the supernate into test tube #2 and save for Step 2.
   (c) Add 5 drops of dilute hydrochloric acid, HCl, into test tube #1 and stir thoroughly. Clean a flame test wire with hydrochloric acid and dip it into the solution. Place the wire loop in a hot flame and record the color.

2. *Separation and Identification of Ca^{2+} in an Unknown Solution*
   (a) Add 10 drops of ammonium oxalate, $(NH_4)_2C_2O_4$, to the solution in test tube #2.

   NOTE: If there is no precipitate, Ca^{2+} is absent. Go directly to Step 3. *↳ yes ppt*

   (b) Centrifuge and then test for completeness of precipitation by adding another drop of ammonium oxalate. Pour off the supernate into test tube #3 and save for Step 3.
   (c) Add 5 drops of dilute hydrochloric acid into test tube #2 and stir thoroughly. Clean a flame test wire with dilute HCl and dip the wire into the solution. Place the wire loop in a hot flame and record the color.

   NOTE: A brick-red flame confirms Ca^{2+} is present.

3. *Identification of Mg^{2+} in an Unknown Solution*
   (a) Add 10 drops of sodium monohydrogen phosphate, Na_2HPO_4, to the solution in test tube #3. Add sodium hydroxide, NaOH, and stir.

   NOTE: If there is no precipitate, Mg^{2+} is absent. *– No ppt*

   (b) Centrifuge and discard the supernate.
   (c) Dissolve the precipitate with dilute hydrochloric acid in test tube #3. Add a few drops of magnesium indicator. Add sodium hydroxide, NaOH, dropwise until the solution tests basic on red litmus paper. Centrifuge.

4. Based upon the observations in steps 1-3, identify the cation(s) present in the unknown solution. *Ca + Ba are present*

PRELABORATORY ASSIGNMENT*

1. In your own words define the following terms: cation, centrifuging, decanting, flame test, precipitate, semimicro analysis, supernate.
2. Why is it necessary to use distilled water throughout this experiment?
3. How do you test for completeness of precipitation?
4. How can litmus paper be used to test for a basic solution?
5. What precautions should be taken while performing this experiment?
6. A cation solution is analyzed for barium, calcium, and magnesium ions. The solution plus $(NH_4)_2SO_4$ gives a white precipitate. The supernate is poured into test tube #2. The precipitate plus HCl gives a green flame test for about two seconds.

    Test tube #2 plus $(NH_4)_2C_2O_4$ gives a white precipitate. The supernate is decanted into test tube #3. The precipitate plus HCl produces a reddish-orange flame test.

    Test tube #3 plus Na_2HPO_4 and NaOH gives a white precipitate. The precipitate dissolves in HCl and magnesium indicator is added. The solution is made basic with NaOH and centrifuged. A clear blue jel is seen at the bottom of the test tube.

    Refer to Figure 9-4 and determine which of these cations are present: Ba^{2+}, Ca^{2+}, Mg^{2+}.

*Answers in Appendix II.

Anion Analysis

OBJECTIVES

1. To observe the chemical behavior of iodide, chloride, and sulfate ions.
2. To analyze an unknown solution for one or more of the following anions: I^{1-}, Cl^{1-}, SO_4^{2-}.
3. To develop the following laboratory skills: centrifuging, washing a precipitate, and using litmus paper.

DISCUSSION

The analysis of a solution for the anions present is based upon the separation and identification of each ion. Separation is achieved by selecting a reagent that reacts differently with the various ions in solution. For example, a sulfate ion will not react with silver nitrate. However, the iodide and chloride ions form a silver precipitate.

When a chloride anion and a silver cation are together in the same solution, a precipitate will form because silver chloride is insoluble. Iodide will also precipitate because silver iodide is insoluble. However, sulfate does not precipitate because silver sulfate is soluble. Figure 10-1 illustrates the separation of I^{1-} and Cl^{1-} from SO_4^{2-}. The separation takes place by centrifuging the solution. The supernate is then poured off and the analysis continues.

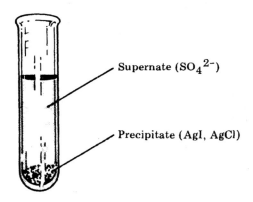

FIGURE 10-1   Test tube after precipitating I^{1-}, and Cl^{1-} and centrifuging the known solution.

In this experiment you will separate and identify I^{1-}, Cl^{1-}, and SO_4^{2-}. First, a known solution containing all three anions will be analyzed to develop techniques and observe reactions. Second, an unknown solution with one or more of the three anions will be analyzed to determine the ions present.

A semimicro approach is employed in this experiment. That is, small test tubes and drop amounts of reagents will be used. Precipitates will be separated by centrifuging. Centrifuging forces the solid particles to the bottom of the test tube where they pack firmly. This allows the supernate to be poured off free of precipitate particles.

Litmus paper is used to determine whether a solution is acidic or basic. A stirring rod is placed into the solution and touched to the litmus paper. Acidic solutions turn blue litmus paper red (Figure 10-2). Basic solutions turn red litmus paper blue.

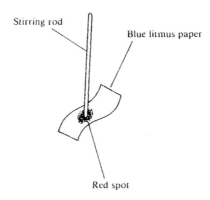

FIGURE 10-2   Testing an acid using blue litmus paper. A red spot is produced.

We will begin the analysis with a known solution containing I^{1-}, Cl^{1-}, and SO_4^{2-}. Silver nitrate is added and the separation begins. Figure 10-3 presents an overview of the analysis. In Step 1, I^{1-} is confirmed. Step 2 confirms Cl^{1-} and Step 3 confirms SO_4^{2-}.

EQUIPMENT AND CHEMICALS

Equipment

- 13 × 100 mm test tubes (3)
- thin glass stirring rod
- wash bottle with distilled water
- centrifuge
- blue litmus paper

- silver nitrate solution 0.2 M $AgNO_3$
- barium nitrate solution, 0.2 M $Ba(NO_3)_2$
- dilute ammonia water, 6 M $NH_3 \cdot H_2O$
- dilute nitric acid, 6 M HNO_3
- unknown solutions (containing one or more of the above anions in 0.1 M concentration)

Chemicals

- known solution containing I^{1-}, Cl^{1-}, and SO_4^{2-} (0.1 M NaI, NaCl, Na_2SO_4)

PROCEDURE

General Directions   Clean three test tubes and a stirring rod with distilled water. Label the test tubes #1, #2, and #3. As a solution is analyzed, record the color of each precipitate in the Data Table.

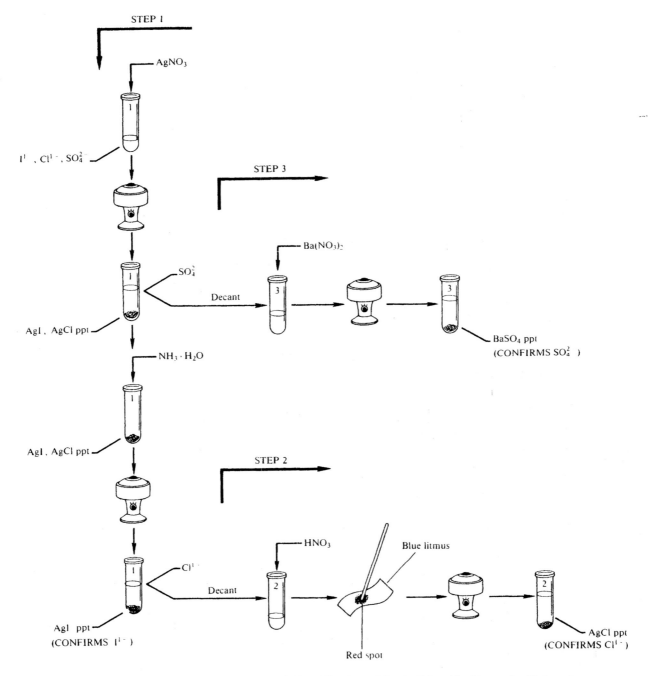

FIGURE 10-3 Separation and identification of the iodide, chloride, and sulfate anions.

A. Analysis of a Known Solution

1. *Separation and Identification of I^{1-} in a Known Solution of I^{1-}, Cl^{1-}, and SO_4^{2-}*
   (a) Place 10 drops of the known solution into test tube #1. Add 20 drops of silver nitrate, $AgNO_3$, and mix with a stir rod.

   NOTE: A yellow precipitate, AgI, suggests I^{1-} is present. AgCl is a white precipitate.

   (b) Centrifuge and decant the supernate into test tube #3 and save for Step 3.
   (c) Add 10 drops of dilute ammonia water, $NH_3 \cdot H_2O$, into test tube #1 and mix thoroughly with a stir rod. Centrifuge and decant the supernate into test tube #2 and save for Step 2. Wash the precipitate.

   NOTE: A yellow precipitate, AgI, confirms I^{1-} is present.

2. *Identification of Cl^{1-} in a Known Solution.* Add dilute nitric acid, HNO_3, dropwise into test tube #2 until the solution tests acidic on blue litmus paper. Centrifuge.

   NOTE: A white precipitate, AgCl, confirms Cl^{1-} is present.

3. *Identification of SO_4^{2-} in a Known Solution.* Add 10 drops of barium nitrate, $Ba(NO_3)_2$, to the solution in test tube #3. Centrifuge.

   NOTE: A white precipitate, $BaSO_4$, confirms SO_4^{2-} is present.

B. Analysis of an Unknown Solution

1. *Separation and Identification of I^{1-} in an Unknown Solution*
   (a) Place 10 drops of unknown solution into test tube #1. Add 20 drops of silver nitrate, $AgNO_3$, and mix with a stir rod.

   NOTE: If there is no precipitate, I^{1-} and Cl^{1-} are absent. Go directly to step 3.

   (b) Centrifuge and decant the supernate into test tube #3 and save for Step 3.
   (c) Add 10 drops of dilute ammonia water, $NH_3 \cdot H_2O$, into test tube #1 and mix thoroughly with a stir rod. Centrifuge and decant the supernate into test tube #2 and save for Step 2. Wash the precipitate.

   NOTE: If there is no precipitate, I^{1-} is absent. Go directly to Step 2.

2. *Identification of Cl^{1-}.* Add dilute nitric acid, HNO_3, dropwise into test tube #2 until the solution tests acidic on blue litmus paper. Centrifuge.

   NOTE: If there is no precipitate, Cl^{1-} is absent. Go directly to Step 3.

3. *Identification of SO_4^{2-}.* Add 10 drops of barium nitrate, $Ba(NO_3)_2$, to the solution in test tube #3. Centrifuge.

   NOTE: If there is no precipitate, SO_4^{2-} is absent.

4. Based upon the observations in steps 1-3, identify the anion(s) present in the unknown solution.

PRELABORATORY ASSIGNMENT*

1. In your own words define the following terms: anion, centrifuging, decanting, precipitate, semimicro analysis, supernate.
2. Why is it necessary to use distilled water throughout the experiment?
3. How do you wash a precipitate?
4. How is litmus paper used to test for an acidic solution?
5. What precautions should be taken while performing the experiment?
6. An anion solution is analyzed for iodide, chloride, and sulfate ions. The solution plus $AgNO_3$ gives a yellow precipitate. The supernate is poured into test tube #3. The yellow precipitate does not dissolve completely in $NH_3 \cdot H_2O$. The supernate is decanted into test tube #2.

   Test tube #2 plus HNO_3 produces a white precipitate. Test tube #3 plus $Ba(NO_3)_2$ yields a white precipitate.

   Refer to Figure 10-3 and determine which of these anions are present in the original solution: I^{1-}, Cl^{1-}, SO_4^{2-}.

* Answers in Appendix II.

DATA TABLE FOR ANION ANALYSIS

A. Analysis of a Known Solution

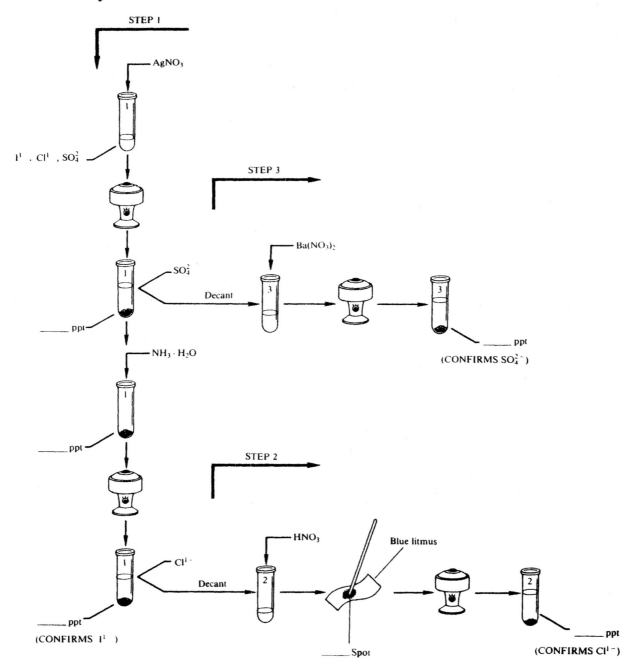

STEP 1

—— AgNO$_3$

I^1 , Cl^1 , SO_4^2

STEP 3

—— SO_4^2

Decant

—— Ba(NO$_3$)$_2$

—— ppt

—— ppt
(CONFIRMS SO_4^{2-})

—— NH$_3 \cdot$ H$_2$O

—— ppt

STEP 2

—— Cl^{1-}

Decant

—— HNO$_3$

Blue litmus

—— ppt
(CONFIRMS I^1)

—— Spot

—— ppt
(CONFIRMS Cl^{1-})

107

B. Analysis of an Unknown Solution

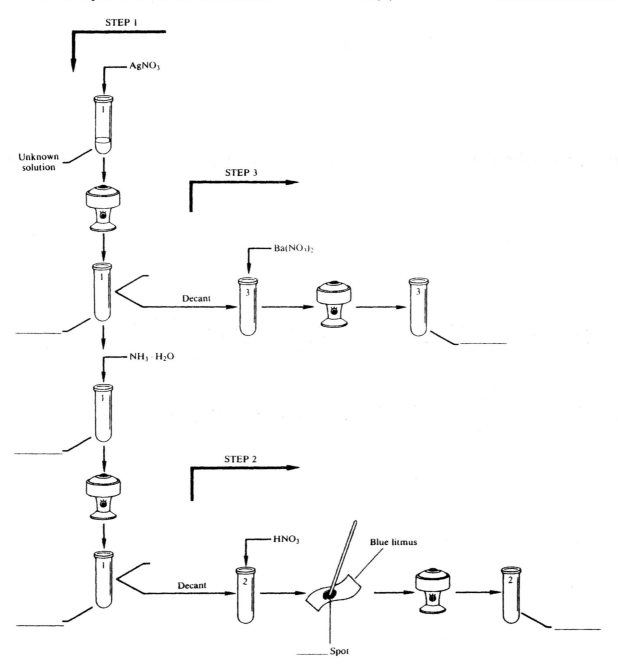

1. An iodide/chloride/sulfate unknown solution gave a yellow precipitate with $AgNO_3$. The supernate did not react with $Ba(NO_3)_2$. The yellow precipitate did not dissolve in $NH_3 \cdot H_2O$. After centrifuging the supernate was acidified with HNO_3 and gave a white precipitate. State the ion(s) present in the unknown solution.

_____

2. An iodide/chloride/sulfate unknown solution plus $AgNO_3$ gave no reaction. The solution was treated with $Ba(NO_3)_2$ and produced a white precipitate. State the ion(s) present in the unknown solution.

_____

3. Write the formula for each of the following anions. (Refer to Appendix XI as necessary.)

(a) acetate   _____   (b) carbonate   _____

(c) chlorite   _____   (d) hydrogen sulfate   _____

(e) hydroxide   _____   (f) perchlorate   _____

(g) nitride   _____   (h) chromate   _____

(i) oxalate   _____   (j) sulfide   _____

4. Name each of the following anions.

(a) F^{1-} _____   (b) CN^{1-} _____

(c) ClO_3^{1-} _____   (d) $Cr_2O_7^{2-}$ _____

(e) HCO_3^{1-} _____   (f) Br^{1-} _____

(g) ClO^{1-} _____   (h) NO_2^{1-} _____

(i) MnO_4^{1-} _____   (j) HSO_3^{1-} _____

5. Complete the table below as shown by the example. Combine the ions into a correct formula and name the compound.

	Br^{1-}	O^{2-}	P^{3-}
K^{1+}	KBr potassium bromide		
Zn^{2+}			
Bi^{3+}			

6. (optional) Complete the table below as shown by the example. Write the formula for each ion and combine the ions into a formula of a compound.

	nitrate	sulfite	phosphate
cuprous	Cu^{1+} NO_3^-		
nickel (II)			
mercurous			
lead (IV)			
ferric			

Avogadro's Number

OBJECTIVES

1. To determine a value of Avogadro's number using a monolayer technique.
2. To find the number of molecules in a monolayer.
3. To find the moles of stearic acid in a monolayer.
4. To develop sensitive techniques in preparing a thin film of molecules.

DISCUSSION

A mole of any substance contains an extremely large number of atoms or molecules. A *mole* is formally defined as the amount of substance containing the same number of particles as there are atoms in exactly 12 grams of carbon-12. This number has been found experimentally and is referred to as *Avogadro's number*.

Avogadro's number *(N)* has been determined by several methods. The most precise value presently available is 6.0221367×10^{23}. However, the most purposes three significant digits are satisfactory.

$$1 \text{ MOLE (mol)} = N = 6.02 \times 10^{23} \text{ particles}$$

The mass of one mole of carbon-12 is 12 grams. We learned earlier that the atomic mass of all other elements is proportional to carbon-12. Therefore, one mole of any other element has a mass equal to its atomic mass expressed in grams. One mole of sodium atoms weighs 23.0 grams, one mole of magnesium atoms weighs 24.3 grams, one mole of oxygen molecules weighs 32.0 grams, and so on. Since we know the mass of one mole of any substance we also know the number of atoms corresponding to the gram-atomic mass, or molecules in a gram-molecular mass.

$$6.02 \times 10^{23} \text{ atoms} = 1 \text{ MOLE} = \text{gram-atomic mass}$$
$$6.02 \times 10^{23} \text{ molecules} = 1 \text{ MOLE} = \text{gram-molecular mass}$$
$$6.02 \times 10^{23} \text{ formula units} = 1 \text{ MOLE} = \text{gram-formula mass}$$

In this experiment we will determine the value of Avogadro's number. First a solution of stearic acid is prepared by dissolving the solid substance in the organic solvent hexane. The solution is added drop-wise onto the water contained in a watchglass. After each drop is added, the stearic acid molecules spread across the surface of the water forming a single layer. This single layer of molecules is referred to as a monolayer. A few seconds after each drop of solution is added the hexane solvent evaporates and the drop disappears. When enough drops of solution have been added to form a monolayer of stearic acid molecules, *one* additional drop forms a clear bead or lens (Figure 11-1).

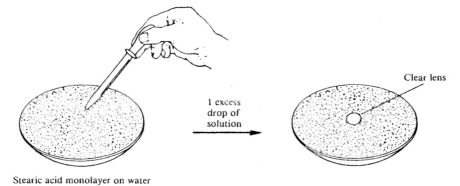

Stearic acid monolayer on water

FIGURE 11-1   A single layer of stearic acid molecules is spread on the surface of water contained in a watchglass.

The stearic acid molecule is a long-chain molecule having a polar "head" and a nonpolar "tail."

$$CH_3-CH_2-CH_2-CH_2-CH_2-CH_2-CH_2-CH_2-CH_2-CH_2-CH_2-CH_2-CH_2-CH_2-CH_2-CH_2-CH_2-COOH$$

Nonpolar "tail"   Polar "head"

The polar "head" is soluble in water which is also polar. The nonpolar "tail" is insoluble in water, just as oil and water are insoluble. The monolayer is composed of stearic acid molecules which have their polar "heads" dissolved in the water and their nonpolar "tails" repelled away from the surface of the water (Figure 11-2).

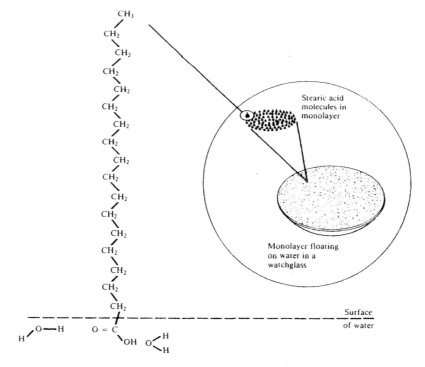

FIGURE 11-2   Enlarged view of the stearic acid molecule in the monolayer.

The cross-sectional area of the stearic acid molecule has been determined from X-ray diffraction patterns. The area occupied by one stearic acid molecule on the surface of the water is 21 Å^2. One Angstrom (Å) is a unit of length equal to 1×10^{-8} cm.

PROBLEM EXAMPLE 11-1

Eleven drops of stearic acid solution produced a circular monolayer with a diameter of 14.5 cm. Calculate the number of molecules in the monolayer assuming each stearic acid molecule occupies an area of 21 Å^2.

Solution: First, we must find the surface area of all the molecules in the monolayer. The formula for a circular surface area may be written

$$\frac{\pi \, d^2}{4} = \text{surface area}$$

$$\frac{(3.14)\,(14.5 \text{ cm})^2}{4} = 165 \text{ cm}^2$$

Second, we simply divide the total surface area by the area of a single molecule. The surface area of the monolayer must first be converted into similar units; that is, Å^2.

$$\frac{165 \text{ cm}^2}{\text{monolayer}} \times \left(\frac{1 \text{ Å}}{1 \times 10^{-8} \text{ cm}} \right)^2 \times \frac{1 \text{ molecule}}{21 \text{ Å}^2}$$

$$= \frac{165 \text{ cm}^2}{\text{monolayer}} \times \frac{1 \text{ Å}^2}{1 \times 10^{-16} \text{ cm}^2} \times \frac{1 \text{ molecule}}{21 \text{ Å}^2}$$

$$= 7.9 \times 10^{16} \text{ molecules/monolayer}$$

In this example the monolayer contains 7.9×10^{16} molecules of stearic acid.

Next, we can calculate the number of moles of stearic acid in the monolayer. However, we must know (1) the concentration of the stearic acid, and (2) the drops of solution that were delivered by the dropper pipet. Let's continue with the determination of Avogadro's number that we began in the preceding problem example.

PROBLEM EXAMPLE 11-2

The monolayer required 11 drops of solution from a pipet calibrated to deliver 75 drops/mL. The concentration of the stearic acid solution is 1.42×10^{-4} g/mL. Calculate the moles of stearic acid in the monolayer. (The gram-molecular mass of stearic acid is 284 g/mol.)

Solution: The moles of stearic acid in the monolayer are calcuated as follows:

$$\frac{11 \text{ drops}}{\text{monolayer}} \times \frac{1 \text{ mL solution}}{75 \text{ drops}} \times \frac{1.42 \times 10^{-4} \text{ g}}{1 \text{ mL solution}} \times \frac{1 \text{ mol}}{284 \text{ g}}$$

$$= 7.3 \times 10^{-8} \text{ mol/monolayer}$$

Thus, the monolayer contains 7.3×10^{-8} moles of stearic acid molecules.

Finally, we can arrive at a value for Avogadro's number by comparing the molecules of stearic acid to the moles of stearic acid in the monolayer.

$$\text{Avogadro's number} = \frac{7.9 \times 10^{16} \text{ molecules}}{7.3 \times 10^{-8} \text{ mol}}$$

$$N = 11 \times 10^{23} \text{ molecules/mol}$$

Notice that we did not express the experimental value for N in scientific notation. In order to better appreciate the result we used 10^{23}, the same exponent as the theoretical value. This value is typical of the results that can be expected from the experiment.

EQUIPMENT AND CHEMICALS

- dropper pipet or 15 cm of 6 mm glass tubing + bulb
- 16 X 150 mm test tube
- 10 mL graduated cylinder
- 15 cm watchglass
- wash bottle with distilled water
- metric ruler
- stearic acid solution (A concentration of 0.12-0.15 g stearic acid/L of hexane is recommended.)

PROCEDURE

A. Calibrating a Dropper Pipet

1. Cut a 15 cm length of 6 mm (OD) soft glass tubing. Heat the tubing and draw it into a fine capillary tip. (Refer to Appendix IV).

   NOTE: A commercially available fine-tip dropper pipet may be substituted. However, for best results the dropper calibration should exceed 70 drops/mL.

2. Obtain about 2 mL of stearic acid solution in a test tube. Calibrate the dropper pipet by adding drops of stearic acid solution into a 10 mL graduated cylinder. Hold the dropper at a 45° angle and deliver the drops at a rate of one per second. Record the number of drops to reach the 1 mL mark.
3. Repeat the calibration procedure twice and find the average number of drops for the three trials.

B. Determining Avogadro's Number

1. Scrupulously clean a large watchglass with soap and water. Rinse the watchglass thoroughly with distilled water and do not touch the inside concave surface.
2. Place the convex side of the watchglass on a paper towel on the lab bench. Fill the watchglass with distilled water from a wash bottle.
3. Record the concentration of the stearic acid solution. Hold the dropper pipet at 45° and slowly deliver drops of stearic acid solution onto the center of the water surface. Observe the drop spreading across the surface. The hexane solvent will evaporate in a few seconds. Continue adding a drop of solution every few seconds until a clear lens persists for at least 30 seconds. The clear lens indicates a monolayer has formed across the entire surface area of the water. Record the number of drops required to form the monolayer.
4. Measure the diameter of the monolayer to the nearest 0.1 cm. Record the diameter and calculate the surface area of the monolayer.
5. Thoroughly clean and rinse the watchglass and perform a second and third trial.
6. Calculate the molecules of stearic acid in the monolayer. Find the moles of stearic acid in the monolayer.
7. Determine the experimental value for Avogadro's number.

PRELABORATORY ASSIGNMENT*

1. In your own words define the following terms: Angstrom unit, Avogadro's number, gram-molecular mass, mole, monolayer, surface area.

2. The stearic acid molecule may be represented as

$CH_3-CH_2-CH_2-CH_2-CH_2-CH_2-CH_2-CH_2-CH_2-CH_2-CH_2-CH_2-CH_2-CH_2-CH_2-CH_2-CH_2-COOH$

   Circle the end of the molecule that is polar and dissolves in water.

3. What is the surface area of a circular monolayer having a diameter of 15.0 cm? What is the area occupied by a single stearic acid molecule in the monolayer? How many stearic acid molecules are present in the monolayer?

4. What is observed after the hexane solvent evaporates?

5. When do you stop adding drops of stearic acid solution to the monolayer?

6. What are the sources of error in this experiment?

7. What safety precautions should be observed in this experiment?

*Answers in Appendix II.

DATA TABLE FOR AVOGADRO'S NUMBER

A. Calibrating a Dropper Pipet

number of drops of solution   _____   _____   _____

Average value   _____ drops/mL

B. Determining Avogadro's Number

concentration of stearic
acid solution   _____ g/mL

number of drops in monolayer   _____   _____   _____

diameter of monolayer   _____ cm   _____ cm   _____ cm

surface area of stearic
acid molecule   _____ $Å^2$ _____ $Å^2$ _____ $Å^2$

Show the calculation for the surface area of the monolayer for trial 1.

surface area of monolayer   _____ cm^2 _____ cm^2 _____ cm^2

B. Determining Avogadro's Number (cont.)

Show the calculation for the number of molecules in the monolayer for trial 1.

molecules in monolayer   _____   _____   _____

Show the calculation for the moles of stearic acid in the monolayer for trial 1.

moles in monolayer   _____ mol   _____ mol   _____ mol

Show the calculation for the number of molecules of stearic acid per mole for trial 1.

Avogadro's number *(N)*   _____   _____   _____

Average value for *N*   _____ molecules/mol

1. A solution of oleic acid, $C_{17}H_{33}COOH$, is added a drop at a time onto a water surface in a 1000 mL beaker until a monolayer is formed. Assume that the surface area of each oleic acid molecule is 46 Å^2 and that there are no spaces between molecules in the monolayer. Calculate the value of Avogadro's number given the following data:

dropper pipet calibration	= 250 drops
number of drops of oleic acid required to form monolayer	= 16 drops
diameter of monolayer	= 10.0 cm
concentration of oleic acid solution	= 0.125 g/L

$N =$ _____

2. Calculate the number of molecules in 0.325 mol of nitrogen gas, N_2.

_____

3. How many moles of iron contain 3.33×10^{24} atoms?

_____

4. What is the mass of 5.12×10^{-3} mol of calcium hydroxide?

_____

5. Find the number of formula units in 1.37 g of sodium chloride.

_____

6. Compute the gram mass of one molecule of stearic acid, $C_{17}H_{35}COOH$.

_____

7. (optional) The oceans of the world cover 3.61×10^6 square kilometers of the earth's surface. A tennis ball floating on water occupies a surface area of 31.6 square centimeters. Assuming there is no space between tennis balls, how many worlds are required to float Avogadro's number of tennis balls?

_____

Percentage of Water in a Hydrate

OBJECTIVES

1. To determine the percentage of water in barium chloride dihydrate.
2. To determine the percentage of water in an unknown hydrate salt.
3. To calculate the water of crystallization for the unknown hydrate salt.
4. To develop the laboratory skills for analyzing a hydrate.

DISCUSSION

A *hydrate salt* is one in which a fixed number of water molecules is crystallized with each formula unit. The number of associated water molecules is referred to as the *water of crystallization*. For example, barium chloride dihydrate, $BaCl_2 \cdot 2H_2O$, has two waters of crystallization. Other common hydrates have waters of crystallization ranging from one to twelve. Upon heating, a hydrate decomposes and produces an *anhydrous salt* and steam.

$$BaCl_2 \cdot 2H_2O_{(s)} \xrightarrow{\Delta} BaCl_{2(s)} + 2 H_2O_{(g)}$$

| hydrate salt | anhydrous salt | water of crystallization |

The *theoretical percentage* of water in a hydrate is found by comparing the mass of the water of crystallization to the mass of the hydrate salt. This is accomplished after finding the formula mass of the hydrate.

PROBLEM EXAMPLE 12-1

Calculate the theoretical percentage of water in barium chloride dihydrate.

Solution: The formula mass of $BaCl_2 \cdot 2H_2O$ is

$BaCl_2 = 208.3$

$$
\begin{aligned}
1 \times 137.3 &= 137.3 \text{ amu} \\
2 \times 35.5 &= 71.0 \text{ amu} \\
2 \times 18.0 &= \underline{36.0 \text{ amu}} \\
& 244.3 \text{ amu}
\end{aligned}
$$

The theoretical percentage of water is found after dividing the water of crystallization mass (36.0 amu) by the hydrate mass (244.3 amu).

$$\frac{36.0 \text{ amu}}{244.3 \text{ amu}} \times 100 = 14.7\% \text{ water}$$

The *experimental percentage* of water in a hydrate is found by comparing the mass of water driven off to the total mass of the compound, expressed as a percentage.

PROBLEM EXAMPLE 12-2

A 1.250 g sample of barium chloride dihydrate has a mass of 1.060 g after heating. Calculate the experimental percentage of water.

Solution: The mass of water lost as steam is found by difference:

$$1.250 \text{ g} - 1.060 \text{ g} = 0.190 \text{ g}$$

hydrate   anhydrous   water
salt

The experimental percentage of water is

$$\frac{\text{mass of water}}{\text{mass of hydrate}} \times 100 = \% \text{ water}$$

$$\frac{0.190 \text{ g}}{1.250 \text{ g}} \times 100 = 15.2\% \text{ water}$$

The proficiency of your laboratory technique is readily measured by comparing the experimental percentage of water to the theoretical value. In the previous examples the two values are in good agreement. The theoretical percentage of water in $BaCl_2 \cdot 2H_2O$ is 14.7 percent and the experimental value was found to be 15.2 percent.

In this experiment barium chloride dihydrate and an unknown hydrate will be heated to determine the percentage of water in each salt (Figure 12-1).

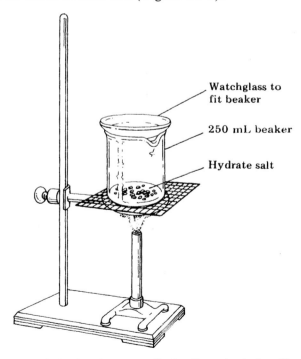

FIGURE 12-1   Apparatus for heating a hydrate salt.

An unknown hydrate will be analyzed using the same procedure as for barium chloride dihydrate. Given the formula mass of the unknown anhydrous salt, it is possible to determine the water of crystallization.

PROBLEM EXAMPLE 12-3

Calculate the water of crystallization for an unknown hydrate salt that is found to contain 30.6 percent water. The formula mass of the anhydrous salt (AS) is 245 amu.

Solution: The unknown hydrate is 30.6 percent water. Subtracting from 100 percent, the hydrate must be 69.4 percent anhydrous salt. If we consider a 100.0 g sample, the mass of water is 30.6 g and the anhydrous salt is 69.4 g. Let's calculate the moles of water and anhydrous salt (AS).

$$30.6 \text{ g } H_2O \times \frac{1 \text{ mole } H_2O}{18.0 \text{ g } H_2O} = 1.70 \text{ moles } H_2O$$

$$69.4 \text{ g AS} \times \frac{1 \text{ mole AS}}{245 \text{ g AS}} = 0.283 \text{ mole AS}$$

To find the water of crystallization we simply divide the mole ratio of water to anhydrous salt.

$$\frac{1.70 \text{ moles } H_2O}{0.283 \text{ mole AS}} = 6.01 \approx 6$$

The water of crystallization is always a whole number; therefore we rounded the ratio to 6. The formula of the unknown hydrate is $AS \cdot 6H_2O$.

EQUIPMENT AND CHEMICALS

Equipment

- wire gauze
- 250 mL beaker
- watchglass

Chemicals

- barium chloride dihydrate,
- $BaCl_2 \cdot 2H_2O$
- unknown hydrate salts

PROCEDURE

A. Percentage of Water in Barium Chloride Dihydrate

1. Weigh a clean, dry 250 mL beaker with watchglass to fit. Add about 1.2 to 1.8 g of barium chloride dihydrate into the beaker and reweigh accurately.
2. Cover the beaker with a watchglass and place on a wire gauze supported on a ring stand. Heat the hydrate gently to avoid spattering the salt. You will observe moisture on the sides of the beaker and the bottom of the watchglass. Continue heating until all the moisture is evaporated. The salt will change from a crystalline to powder form.
3. Turn off the burner and allow the beaker to cool for 10 minutes. Carefully transfer the beaker with the watchglass to the balance. Weigh the mass of the beaker, watchglass and anhydrous salt.
4. Empty the beaker contents into a paper towel and discard. Do a second trial with barium chloride dihydrate. Calculate the percentage of water in the hydrate for each trial and the average value.

B. Percentage of Water in an Unknown Hydrate

1. Obtain an unknown hydrate from the instructor and record the unknown number.
2. Repeat the above procedure and report the average percentage of water in the unknown hydrate.

C. Water of Crystallization in an Unknown Hydrate

1. Given the formula mass of the anhydrous salt from the instructor, calculate the water of crystallization. The instructor may wish to check the experimental percentage of water before giving the formula mass.

   NOTE: It is not unusual for the water of crystallization value to differ by a few tenths from a whole number; for example, 2.1 and 6.7. However, the water of crystallization must be rounded to a whole number; in this example, 2 and 7.

PRELABORATORY ASSIGNMENT*

1. In your own words define the following terms: anhydrous salt, formula mass, hydrate salt, molecular mass, percent composition, water of crystallization, weighing by difference.
2. The mass of a beaker and watchglass is 95.486 g. A hydrate sample is added into the beaker and reweighed. Which of the following indicates a reasonable mass for the beaker, watchglass, and hydrate sample?
   (a) 96.5 g   (b) 96.50 g
   (c) 97.0 g   (d) 96.818 g
   (e) 97.644 g
3. How can you tell when to stop heating because the hydrate is completely decomposed?
4. How will the weighings be affected by placing a warm beaker on the balance?
5. A barium chloride dihydrate sample was analyzed and gave the following data:
   mass of beaker, watchglass + hydrate   102.238 g
   mass of beaker and watchglass   101.046 g
   mass of beaker, watchglass + anhydrous salt   102. 069 g
   Calculate the experimental percentage of water. Does the experimental result agree with the theoretical value calculated in Problem Example 12-1?
6. What safety precautions must be observed in this experiment?

*Answers in Appendix II.

POSTLABORATORY ASSIGNMENT NAME Michael Callanan

1. Calculate the theoretical percentage of water for the following hydrates. Use atomic masses to 0.1 amu and round your answers to three significant digits.

(a) nickel chloride hexahydrate, $NiCl_2 \cdot 6H_2O$

____

(b) calcium nitrate tetrahydrate, $Ca(NO_3)_2 \cdot 4H_2O$

____

(c) zinc sulfate heptahydrate, $ZnSO_4 \cdot 7H_2O$

____

2. A hydrate compound has a mass of 1.632 g before heating and 1.008 g after heating. Compute the experimental percentage of water in the hydrate.

____

3. An unknown salt $AS \cdot XH_2O$ was analyzed and found to contain 16.4 percent water. If the anhydrous salt has a formula mass of 183 amu, find the water of crystallization (X) for the hydrate.

$AS \cdot XH_2O$

4. What is the formula for calcium sulfate hydrate, $CaSO_4 \cdot XH_2O$, containing 20.9 percent water?

    _____

5. A hydrate of cobalt (II) chloride, $CoCl_2$, is found to be 45.5 percent water. Calculate the formula for the hydrate.

    _____

6. The hydrate of chromium (III) nitrate, $Cr(NO_3)_3$, is determined to be 59.5 percent anhydrous salt. Find the formula of the hydrate.

    _____

7. (optional) Find the mass of water driven off when 1.252 g of strontium chloride hexahydrate, $SrCl_2 \cdot 6H_2O$, is decomposed by heating.

    _____

Empirical Formula

OBJECTIVES

1. To determine the empirical formula for magnesium oxide.
2. To determine the empirical formula for copper sulfide.
3. To gain practical experience in developing techniques using a crucible.

DISCUSSION

Empirical formula is defined as the simplest whole number ratio of the elements in a compound. The actual formula for the elements in a compound is termed the *molecular formula* and is the true ratio. For example, hydrogen peroxide has an actual formula of H_2O_2. The simplest ratio of the elements is H_1O_1 and therefore is the empirical formula. Acetylene is a gas used in welding and benzene is a liquid solvent. They have different physical and chemical properties, although both have the same empirical formula, C_1H_1. The molecular formula of acetylene is C_2H_2, whereas the molecular formula of benzene is C_6H_6.

Historically, empirical formulas were determined from the combining weight ratios of elements. This was a critical step in demonstrating the periodic properties of the elements. Empirical formula experiments were also performed in order to determine the combining capacity of the elements. Recently the synthetic element lawrencium was found to have a combining capacity of 3 from an empirical formula experiment. Radioactive lawrencium combined with chlorine to form lawrencium chloride which has the formula $LrCl_3$.

Some elements exhibit more than one combining capacity, and the empirical formula of the compound will depend on how the element combines. For instance, iron can react with oxygen to form either iron(II) oxide or iron(III) oxide, depending on experimental conditions.

PROBLEM EXAMPLE 13-1

A 0.279 g sample of iron was heated and allowed to react with oxygen from the air. The resulting product has a mass of 0.400 g. Find the experimental empirical formula for iron oxide.

Solution: The empirical formula is the whole number ratio of iron and oxygen in the compound iron oxide. This ratio is experimentally determined from the moles of each reactant. The moles of iron are calculated as

$$0.279 \; \text{g Fe} \times \frac{1 \text{ mole Fe}}{55.8 \text{ g Fe}} = 0.00500 \text{ mole Fe}$$

The moles of oxygen is calculated after finding the mass of oxygen reacting.

$$0.400 \text{ g iron oxide} - 0.279 \text{ g Fe} = 0.121 \text{ g O}$$

$$0.121 \; \text{g O} \times \frac{1 \text{ mole O}}{16.0 \text{ g O}} = 0.00756 \text{ mole O}$$

The mole ratio of the elements in iron oxide is

$$Fe_{0.00500} O_{0.00756}$$

We always divide by the smaller number in order to find the simplest whole number ratio:

$$Fe_{\frac{0.00500}{0.00500}} O_{\frac{0.00756}{0.00500}} = Fe_{1.00} O_{1.51}$$

Since we do not have whole numbers it is necessary to double the ratio: $Fe_2 O_{3.02}$. The slight variation from a whole number ratio is accounted for by experimental error. The empirical formula is Fe_2O_3; the compound is iron(III) oxide.

PROBLEM EXAMPLE 13-2

A 1.226 sample of lead shot was placed in a crucible and covered with powdered sulfur. The crucible was heated until all the excess sulfur was driven off. The product weighed 1.417 g. What is the empirical formula of lead sulfide?

Solution: First we will calculate the moles of lead in the product.

$$1.226 \; \text{g Pb} \times \frac{1 \text{ mole Pb}}{207 \text{ g Pb}} = 0.00592 \text{ mole Pb}$$

Second we will calculate the moles of sulfur that combined with the lead.

$$1.417 \text{ g lead sulfide} - 1.226 \text{ g Pb} = 0.191 \text{ g S}$$

$$0.191 \; \text{g S} \times \frac{1 \text{ mole S}}{32.1 \text{ g S}} = 0.00595 \text{ moles S}$$

The mole ratio of the elements in lead sulfide is

$$Pb_{0.00592} S_{0.00595}$$

To simplify we divide by the smaller value:

$$Pb_{\frac{0.00592}{0.00592}} S_{\frac{0.00595}{0.00592}} = Pb_{1.00} S_{1.01}$$

The empirical formula of the product is PbS. The product is lead(II) sulfide.

In this experiment magnesium ribbon will be heated in a crucible and converted to an oxide product. The second part of the experiment involves the conversion of copper to copper sulfide. Since copper can form either copper(I) sulfide or copper(II) sulfide, the empirical formula is unknown and cannot be predicted. Figure 13-1 illustrates the experimental equipment.

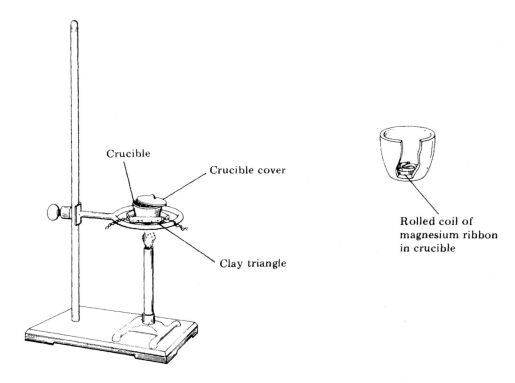

FIGURE 13-1 Apparatus for determining empirical formula.

EQUIPMENT AND CHEMICALS

Equipment

- clay triangle
- crucible tongs
- crucible and cover

Chemicals

- magnesium, Mg ribbon
- copper, #18 gauge Cu wire
- sulfur, S powder

PROCEDURE

A. Empirical Formula of Magnesium Oxide

1. Support a crucible and cover with a clay triangle and place on a ring stand. Fire to red heat.
2. Remove the heat and allow the crucible to cool for 10 minutes. Using crucible tongs, transfer the crucible to the balance and find the mass.
3. Cut a 25 cm strip of magnesium ribbon and roll it into a flat coil. Place the loosely coiled magnesium flat against the bottom of the crucible and reweigh.
4. Return the crucible to the clay triangle. With the lid off, heat the crucible until the magnesium sparks and begins to smoke. Immediately remove the heat and place the cover firmly on the crucible using the tongs.

   NOTE: Safety goggles should be worn when heating the magnesium without the crucible lid.

5. After the smoke has ceased, continue to heat the crucible until the metal is completely converted to a gray-black residue. The progress of conversion can be checked periodically by removing the heat and raising the cover with the tongs.

6. When the metal no longer sparks, remove the heat and allow the crucible to cool for 10 minutes. Add several drops of distilled water with a dropper pipet to the residue.

   NOTE: Some of the magnesium will react with the nitrogen in the air forming magnesium nitride. The addition of water decomposes any magnesium nitride present releasing ammonia gas.

7. Refire to red heat for 5 minutes. Remove the heat and allow the crucible with product to cool for 10 minutes. Transfer the crucible to the balance and find the mass of the crucible and magnesium oxide.

8. Clean the crucible, repeat the procedure, and calculate the empirical formula for each trial.

B. Empirical Formula of Copper Sulfide

1. Support a crucible and cover with a clay triangle and place on a ring stand. Fire to red heat.

2. Remove the heat and allow the crucible to cool for 10 minutes. Using crucible tongs, transfer the crucible to the balance and find the mass.

3. Cut a 25 cm length of copper wire. Roll the wire into a coil, place it in the crucible, and reweigh.

4. Completely cover the copper with powdered sulfur. Place the lid on the crucible and gradually heat to red heat under a fume hood. Continue to heat for several minutes after the last trace of burning sulfur disappears. Holding the burner in your hand, heat the entire outside surface of the crucible and cover.

5. Allow the crucible and contents to cool for 10 minutes. Transfer the crucible to a balance and weigh the crucible, cover, and copper sulfide.

6. Clean the crucible, repeat the procedure, and calculate the empirical formula for each trial.

PRELABORATORY ASSIGNMENT*

1. In your own words define the following terms: empirical formula, firing to red heat, heating to constant weight, molecular formula, weighing by difference.

2. Why are crucible tongs used to transfer the crucible after heating and before weighing?

3. How critical are the suggested times for heating and cooling?

4. Why is distilled water added to the crucible after the initial heating of the magnesium?

5. How can you tell when the reaction is complete for the magnesium?

6. How can you tell when the copper has reacted completely and the excess sulfur is burned off?

7. What are the major sources of error in this experiment?

8. What safety precautions must be observed in this experiment?

*Answers in Appendix II.

DATA TABLE FOR EMPIRICAL FORMULA

A. Empirical Formula of Magnesium Oxide

mass of crucible and cover + magnesium
(before heating)   _____ g   _____ g

mass of crucible and cover   _____ g   _____ g

mass of magnesium   _____ g   _____ g

mass of crucible and cover + magnesium
oxide *(after heating)*   _____ g   _____ g

mass of combined oxygen   _____ g   _____ g

Show the calculation of empirical formula for trial 1.

Empirical formula of magnesium oxide   _____   _____

B. Empirical Formula of Copper Sulfide

mass of crucible and cover + copper
(*before heating*) _____ g _____ g

mass of crucible and cover _____ g _____ g

mass of copper _____ g _____ g

mass of crucible and cover + copper
sulfide (*after heating*) _____ g _____ g

mass of combined sulfur _____ g _____ g

Show the calculation of empirical formula for trial 1.

Empirical formula of copper sulfide _____ _____

1. A 0.750 g sample of tin is oxidized with nitric acid to form tin oxide. Calculate the empirical formula of tin oxide if the original tin sample gained 0.201 g of oxygen.

_____

2. Excess sulfur reacts with 0.565 g of cobalt to give 1.027 g of cobalt sulfide. Find the empirical formula of the product.

_____

3. If 1.164 g of iron filings react with chlorine gas to give 3.384 g of iron chloride, what is the empirical formula of the compound?

_____

4. A 0.626 g sample of copper oxide was reduced to 0.500 g of copper metal by heating in a stream of hydrogen. Calculate the empirical formula of the copper oxide sample.

_____

5. A sample of phosphorus weighing 0.500 g was ignited to phosphorus oxide in a stream of pure oxygen. The product has a mass of 1.145 g. Calculate the empirical formula. Find the molecular formula if a separate experiment yielded a molecular mass of approximately 285 amu for the phosphorus oxide.

Empirical formula _____

Molecular formula _____

6. (optional) Ethylene glycol, the main ingredient in permanent antifreeze, contains 38.7 percent carbon, 9.7 percent hydrogen, and 51.6 percent oxygen. Calculate the empirical and molecular formulas given a molecular mass of approximately 60 amu.

Empirical formula _____

Molecular formula _____

Chemical Reactions

OBJECTIVES

1. To become familiar with the evidence for chemical reaction.
2. To translate word equations into balanced chemical equations.
3. To prepare metallic copper and determine the atomic mass of an unknown metal (*M*).
4. To systematically study five major types of reactions.

DISCUSSION

Ordinary chemical reactions can be considered to be one of five types. The first type is the synthesis of a single compound from two or more substances. This type of reaction is termed *combination*.

$$A + Z \longrightarrow AZ$$

A second type of reaction is termed *decomposition* in which a single compound breaks down into two or more simpler substances, usually by the application of heat.

$$AZ \longrightarrow A + Z$$

A third type of reaction is called *replacement*; here, one element simply displaces another from a compound. The element that is displaced is lower in the electromotive series.

$$A + BZ \longrightarrow AZ + B$$

In a *double replacement* reaction, two substances in solution switch partners; that is, the anion of one substance exchanges with the anion of another compound.

$$AX + BZ \longrightarrow AZ + BX$$

A fifth type of reaction is a *neutralization* reaction; an acid and a base react to form a salt and water. is this the same as combustion reaction

$$HX + BOH \longrightarrow BX + HOH$$

A neutralization reaction is actually a special type of double replacement reaction, where one cation is hydrogen and one anion is hydroxide. The hydrogen in the acid neutralizes the hydroxide in the base to form water. If the formula of water is written as HOH, the equation is more easily balanced.

In this experiment, each of the five types of reactions is performed. The evidence of reaction is carefully observed and recorded. The evidence may include any of the following: (1) a gas is produced; (2) a precipitate is formed; (3) a color change is observed; (4) a temperature change is noted.

There are many symbols used in chemical equations to convey the reaction conditions. Table 14-1 lists some of these.

TABLE 14-1 Symbols in Chemical Equations

Symbol	Translation
$\longrightarrow$	produces, yields (separates reactants from products)
+	added to, reacts with (separates two or more reactants or products)
Δ	heat catalyst (written above $\longrightarrow$)
NR	no reaction (written after $\longrightarrow$)
(s)	solid or precipitate
(ℓ)	liquid
(g)	gas
(aq)	aqueous solution

In order to write an equation it is necessary to predict the products from a given reaction. Initially, this is a difficult task. To aid you in writing equations, word equations are supplied for each reaction. However, it is necessary to translate the word equations into balanced chemical equations. The following examples will illustrate.

EXAMPLE 14-1 COMBINATION REACTION

$$\text{zinc}_{(s)} + \text{oxygen}_{(g)} \xrightarrow{\Delta} \text{zinc oxide}_{(s)}$$

$$2\,Zn_{(s)} + O_{2\,(g)} \xrightarrow{\Delta} 2\,ZnO_{(s)}$$

EXAMPLE 14-2 DECOMPOSITION REACTION

$$\text{nickel(II) chloride hexahydrate}_{(s)} \xrightarrow{\Delta} \text{nickel(II) chloride}_{(s)} + \text{water}_{(g)}$$

$$NiCl_2 \cdot 6H_2O_{(s)} \xrightarrow{\Delta} NiCl_{2\,(s)} + 6\,H_2O_{(g)}$$

EXAMPLE 14-3 REPLACEMENT REACTION

$$\text{tin}_{(s)} + \text{hydrochloric acid}_{(aq)} \longrightarrow \text{tin(II) chloride}_{(aq)} + \text{hydrogen}_{(g)}$$

$$Sn_{(s)} + 2\,HCl_{(aq)} \longrightarrow SnCl_{2\,(aq)} + H_{2\,(g)}$$

EXAMPLE 14-4 DOUBLE REPLACEMENT REACTION

$$\text{potassium carbonate}_{(aq)} + \text{calcium chloride}_{(aq)} \longrightarrow \text{calcium carbonate}_{(s)} + \text{potassium chloride}_{(aq)}$$

$$K_2CO_{3\,(aq)} + CaCl_{2\,(aq)} \longrightarrow CaCO_{3\,(s)} + 2\,KCl_{(aq)}$$

EXAMPLE 14-5 NEUTRALIZATION REACTION

hydrochloric acid$_{(aq)}$ + barium hydroxide$_{(aq)}$ $\longrightarrow$ barium chloride$_{(aq)}$ + water$_{(\ell)}$

$$2 \ HCl_{(aq)} + Ba(OH)_{2\,(aq)} \longrightarrow BaCl_{2\,(aq)} + 2 \ HOH_{(\ell)}$$

Also in this experiment you will prepare metallic copper using an unknown metal. The unknown metal *(M)* displaces copper from a copper sulfate solution as follows:

$$M_{(s)} + CuSO_{4\,(aq)} \longrightarrow MSO_{4\,(aq)} + Cu_{(s)}$$

The mass of solid copper produced is compared to the mass of unknown metal.

PROBLEM EXAMPLE 14-1

A student weighed out 0.367 g of unknown metal *(M)*. After reaction with 25 mL of copper sulfate solution, 0.417 g of metallic copper was collected. Calculate the atomic mass of the unknown metal. The equation for the reaction is shown above.

Solution: From the chemical equation, we see one mole of *M* produces one mole of Cu.

$$0.417 \ g \ Cu \times \frac{1 \ \text{mole Cu}}{63.5 \ g \ Cu} \times \frac{1 \ \text{mole } M}{1 \ \text{mole Cu}} = 0.00657 \ \text{mol } M$$

The atomic mass (g/mole) is:

$$\frac{0.367 \ g \ M}{0.00657 \ \text{mol } M} = 55.9 \ g/mole$$

In this example the atomic mass is found to be 55.9 g/mole, which corresponds to iron. Other unknown metals *(M)* will give a different value.

EQUIPMENT AND CHEMICALS

Equipment

- crucible tongs
- evaporating dish
- 100 mL graduated cylinder
- 250 mL beaker
- wire gauze
- 16 × 150 mm test tubes (6)
- test tube holder
- test tube brush
- 250 mL Erlenmeyer flask
- crucible
- wash bottle

Chemicals

- magnesium, Mg ribbon
- sulfur, S powder
- zinc, Zn powder
- copper(II) sulfate pentahydrate, solid $CuSO_4 \cdot 5H_2O$
- sodium hydrogen carbonate, solid $NaHCO_3$

- wooden splints
- copper, Cu wire
- calcium, Ca turnings
- dilute hydrochloric acid, 6 M HCl
- silver nitrate, 0.1 M $AgNO_3$
- mercury(II) nitrate, 0.1 M $Hg(NO_3)_2$
- aluminum nitrate, 0.1 M $Al(NO_3)_3$
- potassium iodide, 0.1 M KI
- sodium phosphate, 0.1 M Na_3PO_4
- nitric acid, 0.1 M HNO_3
- sulfuric acid, 0.1 M H_2SO_4
- phosphoric acid, 0.1 M H_3PO_4
- sodium hydroxide, 0.1 M NaOH
- phenolphthalein
- copper(II) sulfate, 0.5 M $CuSO_4$
- unknown metal samples

PROCEDURE

NOTE: For Procedures A-E, record your observations in the Data Table. On the page following the observations, word equations are written for each reaction. Write the corresponding balanced chemical equation.

For Procedure F, the unknown metal may require 30 minutes for complete reaction. Therefore, it may be convenient to perform the first two steps of Procedure F before beginning Procedures A-E.

A. Combination Reactions

Do 1 only

1. Hold a 2 cm strip of magnesium ribbon with crucible tongs and ignite the metal in a hot burner flame.
2. Mix together 2 g of powdered zinc and 1 g of sulfur in a crucible and place under a fume hood. Heat the end of a metal wire to red heat and use the end to ignite the mixture.

   CAUTION: The instructor should either demonstrate or closely supervise this reaction. This procedure should be considered dangerous.

B. Decomposition Reactions

Do 1.

1. Add a few crystals of copper(II) sulfate pentahydrate in a dry test tube. Holding the test tube with a test tube holder, heat strongly with a burner. Note the change in color and texture and observe the inside wall of the test tube.
2. Add sodium hydrogen carbonate (baking soda) into a 250 mL Erlenmeyer flask so as to sparsely cover the bottom. Support the flask on a ring stand using a wire gauze.
   (a) Hold a lighted splint inside the neck of the flask and observe how long it burns.
   (b) Heat the flask strongly with the burner and note the inside wall of the flask. After moisture collects, plunge a flaming splint into the flask and observe how long it burns.

C. Replacement Reactions

1. Put 2 mL (about 1/10 test tube) of silver nitrate into a test tube and add a small copper wire. Allow a few minutes for reaction and then record your observation.
2. Place a small piece of magnesium into a test tube containing about 2 mL of dilute hydrochloric acid.
3. Place a small piece of calcium metal into a test tube containing a few milliliters of distilled water.

D. Double Replacement Reactions

1. Put 2 mL of silver nitrate, mercury(II) nitrate, and aluminum nitrate into separate test tubes #1-3. Add about 2 mL of potassium iodide into test tube #1 and check for evidence of reaction.
2. Add about 2 mL of potassium iodide into test tube #2 and check for reaction.
3. Add about 2 mL of potassium iodide into test tube #3 and check for reaction.
4. Put 2 mL of silver nitrate, mercury(II) nitrate, and aluminum nitrate into separate test tubes #4-6. Add about 2 mL of sodium phosphate into test tube #4 and check for evidence of reaction.
5. Add about 2 mL of sodium phosphate into test tube #5 and note the reaction.
6. Add about 2 mL of sodium phosphate into test tube #6 and observe the reaction.

E. Neutralization Reactions

1. Put 2 mL of nitric acid, sulfuric acid, and phosphoric acid into separate test tubes #1-3. Introduce one drop of phenolphthalein indicator into each of the test tubes. Add sodium hydroxide solution to test tube #1 until a permanent color change is observed.

   > NOTE: Phenolphthalein is an acid-base indicator that is colorless in acidic and neutral solutions and pink in basic solutions.

2. Add sodium hydroxide solution into test tube #2 until a permanent color change is observed.
3. Add sodium hydroxide solution into test tube #3 until a permanent color change is observed.

F. Preparation of Copper from an Unknown Metal

1. Weigh a clean, dry evaporating dish. Add 0.2-0.5 g of unknown metal into the dish and reweigh. Find the mass of the metal by difference.
2. Using a graduated cylinder, pour 25 mL of copper(II) sulfate solution into the evaporating dish. Allow the unknown metal to react completely with the copper solution.

   > NOTE: It may be convenient to stop at this step and complete Procedures A-E.

3. After the reaction is complete, carefully pour off the solution so that the copper metal remains in the dish. A small amount of blue solution will remain also.
4. Wash the copper metal with 25 mL of distilled water and discard the washings. Repeat the washing procedure until the copper metal is clear of blue solution.
5. Prepare a waterbath in a 250 mL beaker and place the evaporating dish in the beaker (Figure 14-1). Evaporate the copper metal to dryness.

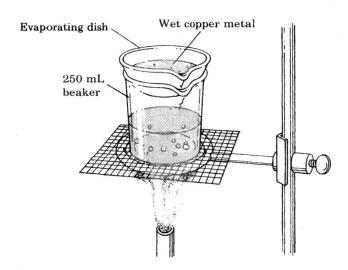

Evaporating dish   Wet copper metal

250 mL
beaker

FIGURE 14-1   Apparatus for drying the metallic copper.

6. After evaporation, remove the dish and wipe dry. Hold the dish with crucible tongs over a flame to remove the last traces of moisture. Allow the dish to cool and weigh the evaporating dish with residue.
7. Calculate the atomic mass (g/mole) of the unknown metal (M). Identify the metal (M) from its atomic mass.

PRELABORATORY ASSIGNMENT*

1. In your own words define the following terms: catalyst, electromotive series, exothermic reaction, precipitate, product, reactant.
2. Explain the meaning of the following symbols:

   $\xrightarrow{\Delta}$, NR, (s), (ℓ), (g), (aq)
3. List four observations that are evidence of chemical reaction.
4. What is the estimated volume of liquid in a test tube that is 1/10 full?
5. What color is the phenolphthalein indicator in an acid solution? In a basic solution?
6. What safety precautions should be followed in this experiment?
7. Calculate the atomic mass of an unknown metal (M) given the following data:

   mass of evaporating dish + unknown metal (M) = 45.882 g
   mass of evaporating dish   = 45.361 g
   mass of evaporating dish + copper metal   = 45.781 g

*Answers in Appendix II.

Analysis by Decomposition

OBJECTIVES

1. To decompose potassium chlorate and determine the percent yield of oxygen gas.
2. To determine the percentage of potassium chlorate in an unknown mixture.
3. To gain proficiency in thermally decomposing a compound and using a gas collection apparatus.

DISCUSSION

If solid potassium chlorate is heated, it decomposes into solid potassium chloride and oxygen gas. Manganese dioxide is used as a catalyst to moderate the rate of decomposition. The equation for the reaction is

$$2 \; KClO_{3(s)} \xrightarrow[\Delta]{MnO_2} 2 \; KCl_{(s)} + 3 \; O_{2(g)}$$

In this experiment we will decompose a potassium chlorate mixture. The mass loss after heating the mixture equals the mass of oxygen gas released. Alternately, we can calculate the mass of oxygen liberated from the stoichiometry of the reaction.

PROBLEM EXAMPLE 15-1

Calculate the percent yield of oxygen gas if 1.850 g of a 90.0% potassium chlorate mixture gives an actual yield of 0.633 g oxygen.

Solution: Since the mixture is 90.0% potassium chlorate, there is only 90.0 g $KClO_3$ in each 100.0 g of mixture.

$$1.850 \; \text{g mixture} \times \frac{90.0 \; \text{g } KClO_3}{100.0 \; \text{g mixture}} = 1.67 \; \text{g } KClO_3$$

The above balanced equation for the reaction relates the mass of potassium chlorate (122.6 amu) to the mass of oxygen (32.0 amu). We can calculate the theoretical yield of oxygen gas as follows:

$$1.67 \; \text{g } KClO_3 \times \frac{1 \; \text{mole } KClO_3}{122.6 \; \text{g } KClO_3} \times \frac{3 \; \text{moles } O_2}{2 \; \text{moles } KClO_3} \times \frac{32.0 \; \text{g } O_2}{1 \; \text{mole } O_2} = 0.654 \; \text{g } O_2$$

The actual yield is given as 0.633 g oxygen; therefore, the percent yield is found as follows:

$$\frac{\text{actual yield}}{\text{theoretical yield}} \times 100 = \% \text{ yield}$$

$$\frac{0.633 \text{ g}}{0.654 \text{ g}} \times 100 = 96.8\%$$

The second part of this experiment analyzes an unknown mixture for its potassium chlorate content. Potassium chlorate in the unknown sample is decomposed as before, releasing oxygen gas. The equation for the reaction is the same as above and the stoichiometry calculations are similar.

PROBLEM EXAMPLE 15-2

A 2.050 g mixture containing potassium chlorate is decomposed by heating. If the mass loss is 0.602 g oxygen, what is the percentage of potassium chlorate in the unknown sample.

Solution: We can relate the mass of oxygen to the mass of potassium chlorate according to the balanced equation. In this example the stoichiometry relationship is indirect as we are relating the mass of the oxygen product to the mass of potassium chlorate reactant.

$$0.602 \text{ g } O_2 \times \frac{1 \text{ mole } O_2}{32.0 \text{ g } O_2} \times \frac{2 \text{ moles } KClO_3}{3 \text{ moles } O_2} \times \frac{122.6 \text{ g } KClO_3}{1 \text{ mole } KClO_3} = 1.54 \text{ g } KClO_3$$

The percentage of $KClO_3$ in the unknown mixture is simply

$$\frac{\text{mass } KClO_3}{\text{mass mixture}} \times 100 = \% \text{ } KClO_3$$

$$\frac{1.54 \text{ g}}{2.050 \text{ g}} \times 100 = 75.1\%$$

Figure 15-1 shows the experimental apparatus for collecting the oxygen gas. As the potassium chlorate decomposes, oxygen gas is produced. The gas displaces water from the Florence flask into a beaker. When the decomposition is complete, no more oxygen is released and the water level in the beaker remains constant. After a few minutes, the water level may actually decrease owing to the cooling of the oxygen gas in the flask.

EQUIPMENT AND CHEMICALS

Equipment

- gas collection apparatus (see Figure 15-1)
- 16 × 150 mm test tube
- 1000 mL Florence flask
- 600 mL beaker
- safety goggles

Chemicals

- potassium chlorate mixture, 90.0% $KClO_3$
- unknown potassium chlorate mixtures (40–90% $KClO_3$)

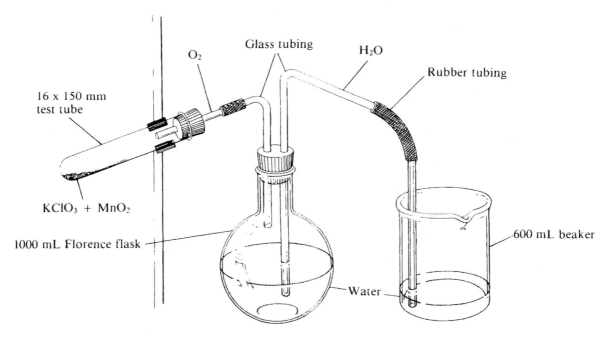

FIGURE 15-1   Decomposition apparatus. When the water level
in the flask remains constant, the decomposition is complete.

PROCEDURE

A. Percent Yield of Oxygen

1. Weigh a 16 × 150 mm test tube on the balance and record the mass. Add about 1.5 g
   potassium chlorate (1–2 g) and reweigh accurately.
2. Set up the apparatus as shown in Figure 15-1. Attach the test tube to a ring stand using
   a utility clamp. Fill the Florence flask with water and insert the rubber stopper with
   glass tubing.

   CAUTION: Do not let any of the mixture come in contact with the rubber stopper in
   the test tube. Ask the instructor to inspect the apparatus before con-
   tinuing.

3. Begin heating the test tube gently. Observe the water being displaced into the beaker
   as oxygen is produced. As the water level in the beaker increases, heat the test tube
   more strongly. After the water level remains constant for a couple of minutes, discon-
   tinue heating and allow the test tube to cool for 10 minutes.
4. Weigh the test tube with contents. The mass of oxygen gas is found by the mass difference before and after heating.
5. Calculate the theoretical yield of oxygen from the mass of the potassium chlorate
   mixture. Find the percent yield.

B. Percentage of Potassium Chlorate in an Unknown Mixture

1. Obtain an unknown sample containing potassium chlorate. Record the unknown number in the Data Table.
2. Repeat Steps 1–3 in the above procedure, substituting the unknown for the 90.0%
   mixture.
3. Calculate the mass of potassium chlorate in the unknown sample from the mass of
   oxygen gas. Find the percentage of potassium chlorate in the unknown mixture.

PRELABORATORY ASSIGNMENT*

1. In your own words, define the following terms: actual yield, mass percent composition, percent yield, stoichiometry, theoretical yield, weighing by difference.
2. Why is manganese dioxide added to potassium chlorate before heating?
3. How do you determine the mass of oxygen gas produced?
4. How can you tell when the sample is completely decomposed?
5. What are the primary sources of error in this experiment?
6. Is it possible to have a percent yield that is greater than 100%?
7. What safety precautions must be observed in this experiment?

*Answers in Appendix II.

NAME _____

DATE _____

SECTION _____

DATA TABLE FOR ANALYSIS BY DECOMPOSITION

A. Percent Yield of Oxygen

mass of test tube + 90.0% mixture
(before heating)   _____ g   _____ g

mass of test tube   _____ g   _____ g

mass of 90.0% mixture   _____ g   _____ g

mass of test tube + residue
(after heating)   _____ g   _____ g

mass of oxygen (actual yield)   _____ g   _____ g

Show the calculation for the theoretical yield of oxygen.

mass of O_2 gas (theoretical yield)   _____ g   _____ g

Show the calculation for percent yield of oxygen.

Percent yield of O_2 _____ %   _____ %

B. Percentage of Potassium Chlorate in an Unknown Mixture

UNKNOWN # _____

mass of test tube + unknown mixture
(*before heating*) _____ g _____ g

mass of test tube _____ g _____ g

mass of unknown mixture _____ g _____ g

mass of test tube + residue
(*after heating*) _____ g _____ g

mass of oxygen _____ g _____ g

Show the calculation for the mass of potassium chlorate in the unknown mixture.

mass of $KClO_3$ _____ g _____ g

Show the calculation for percentage of potassium chlorate in the unknown mixture.

Percentage of $KClO_3$ _____ % _____ %

1. A 1.915 g sample of sodium nitrate mixture (95.0% purity) is decomposed by heating and found to have a resulting mass of 1.582 g. Calculate the percent yield of oxygen.

$$2 \, NaNO_{3(s)} \xrightarrow{\Delta} 2 \, NaNO_{2(s)} + O_{2(g)}$$

2. A 1.634 g sample of sodium nitrate mixture is decomposed by heating. If the mass loss is 0.223 g, what is the percentage of sodium nitrate in the unknown mixture?

3. If a 1.75 g sample of ammonium dichromate is decomposed, what is the milliliter volume of nitrogen gas released at STP? The equation for the reaction is

$$(NH_4)_2Cr_2O_{7(s)} \xrightarrow{\Delta} Cr_2O_{3(s)} + N_{2(g)} + 4\,H_2O_{(l)}$$

4. Magnesium nitride decomposes in water and releases ammonia gas. If 335 mL of ammonia is produced at STP, what is the mass of magnesium nitride that reacted? The equation for the reaction is

$$Mg_3N_{2(s)} + 6\,H_2O_{(l)} \longrightarrow 3\,Mg(OH)_{2(s)} + 2\,NH_{3(g)}$$

5. (optional) A 1.00 g sample mixture containing baking soda is decomposed. If 73.5 mL of carbon dioxide gas is collected over water at 24°C and a partial pressure of 745 mm Hg, what is the percentage of NaHCO$_3$ in the sample mixture? The equation for the reaction is

$$2\,NaHCO_{3(s)} \xrightarrow{\Delta} Na_2CO_{3(s)} + H_2O_{(l)} + CO_{2(g)}$$

Analysis by Precipitation

OBJECTIVES

1. To precipitate potassium iodide with lead(II) nitrate and determine the percent yield of lead(II) iodide.
2. To determine the percentage of potassium iodide in an unknown mixture.
3. To gain proficiency in transferring and filtering a precipitate.

DISCUSSION

In this experiment potassium iodide will be precipitated from aqueous solution with lead(II) nitrate. The equation for the reaaction is:

$$2 \ KI_{(aq)} + Pb(NO_3)_{2(aq)} \longrightarrow PbI_{2(s)} + 2 \ KNO_{3(aq)}$$

The insoluble lead(II) iodide solid is collected in filter paper. The mass of the precipitate is referred to as the *actual yield* of lead(II) iodide. The *theoretical yield* of lead(II) iodide is the calculated amount of precipitate formed according to the stoichiometry of the reaction. It is assumed that all of the potassium iodide reactant has been converted to lead(II) iodide product. The *percent yield* is found by comparing the actual yield to the theoretical yield and expressing the ratio as a percentage.

PROBLEM EXAMPLE 16-1

Calculate the percent yield of lead(II) iodide if 0.995 g of potassium iodide gives an actual yield of 1.141 g precipitate.

Solution: The above balanced equation for the reaction relates the mass of reactant to the precipitate. Given the formula mass of potassium iodide (166.0 amu) and lead(II) iodide (461.0 amu), we can calculate the theoretical yield of precipitate as follows:

$$0.995 \ \text{g KI} \times \frac{1 \ \text{mole KI}}{166.0 \ \text{g KI}} \times \frac{1 \ \text{mole PbI}_2}{2 \ \text{moles KI}} \times \frac{461.0 \ \text{g PbI}_2}{1 \ \text{mole PbI}_2} = 1.38 \ \text{g PbI}_2$$

The actual yield is given as 1.141 g PbI_2; therefore, the percent yield is found as follows:

$$\frac{\text{actual yield}}{\text{theoretical yield}} \times 100 = \% \text{ yield}$$

$$\frac{1.141 \text{ g}}{1.38 \text{ g}} \times 100 = 82.6\%$$

The second part of this experiment analyzes an unknown mixture for its potassium iodide content. Potassium iodide in the unknown sample is precipitated as before, forming insoluble lead(II) iodide. The equation for the reaction is the same as above and the stoichiometry calculations are similar.

PROBLEM EXAMPLE 16-2

Calculate the percentage of potassium iodide in a 1.005 g unknown mixture that produced a 0.915 g precipitate of lead(II) iodide.

Solution: As before, the mass of precipitate can be related to the mass of potassium iodide in the balanced equation. In this example the relationship is indirect as we relate the mass of product back to the mass of potassium iodide reactant.

$$0.915 \text{ g } PbI_2 \times \frac{1 \text{ mole } PbI_2}{461.0 \text{ g } PbI_2} \times \frac{2 \text{ moles } KI}{1 \text{ mole } PbI_2} \times \frac{166.0 \text{ g } KI}{1 \text{ mole } KI} = 0.659 \text{ g } KI$$

The percentage of KI in the unknown mixture is simply

$$\frac{\text{mass KI}}{\text{mass mixture}} \times 100 = \% \text{ yield}$$

$$\frac{0.659 \text{ g}}{1.005 \text{ g}} \times 100 = 65.5\%$$

Figure 16-1a shows the filtering apparatus before transferring the precipitate. Notice the lead(II) iodide has been allowed to settle out of solution before decanting the supernate. If the solution is transferred without allowing the precipitate to settle, the particles will prematurely clog the pores in the filter paper. This will slow the filtering process. Figure 16-1b illustrates washing the precipitate from the beaker into the filter paper.

EQUIPMENT AND CHEMICALS

Equipment

- 250 mL beaker
- 100 mL graduated cylinder
- wire gauze
- clay triangle
- funnel
- filter paper
- 400 mL beaker
- stirring rod
- rubber policeman
- wash bottle

Chemicals

- potassium iodide, anhydrous KI
- lead(II) nitrate solution, 0.1 M $Pb(NO_3)_2$
- unknown potassium iodide mixtures

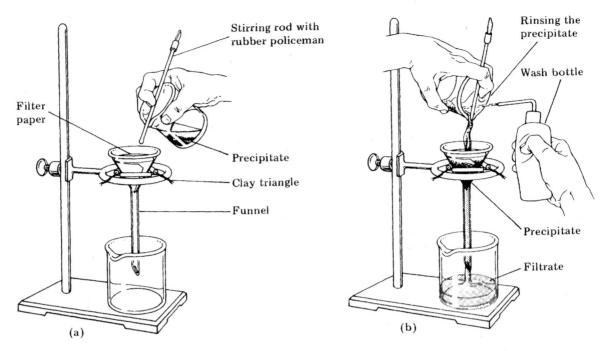

FIGURE 16-1 Filtering apparatus. (a) Allow the precipitate to settle, then pour off the supernate. (b) After the supernate passes through the filter paper, rinse the precipitate into the filter paper using a wash bottle.

PROCEDURE

A. Percent Yield of Lead(II) Iodide

filter paper = 1.13g

1. Place a 250 mL beaker on the balance and record the mass. Add about 1 g potassium iodide (0.8-1.2 g) and reweigh accurately. *109.25 g - Beaker + K1g = 110.45g*
2. Dissolve the sample completely in 25 mL of distilled water. Using a graduated cylinder, transfer 50 mL of lead(II) nitrate solution into the 250 mL beaker.
3. Support the beaker with a wire gauze on a ring stand. Bring the solution to a gentle boil and then turn off the burner. Allow the precipitate to digest until the solution is cool.
4. Weigh a disk of filter paper. Prepare a filter paper cone as shown in Figure 16-2. Insert the filter paper cone into the funnel and moisten with water.

(1) Fold and crease lightly.

(3) Seal the moistened edge of the filter paper against the funnel.

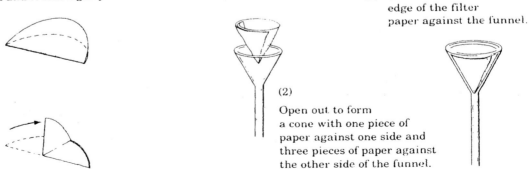

(2)

Open out to form a cone with one piece of paper against one side and three pieces of paper against the other side of the funnel.

FIGURE 16-2 Folding filter paper and inserting it into a funnel.

5. Assemble a filtering apparatus as shown in Figure 16-1.

6. Without disturbing the settled precipitate, carefully pour off the supernate into the filter paper. Use a stirring rod to guide the flow. Rinse out the bulk of the precipitate with a stream of water from the wash bottle. Clean the beaker using a rubber policeman and rinse the residue into the filter paper.

   NOTE: At no time add supernate above the top of the filter paper as precipitate particles can pass into the filtrate.

7. After all of the supernate has passed through the filter, carefully remove the paper cone from the funnel. After drying, weigh the filter paper with precipitate.

8. Calculate the theoretical yield of lead(II) iodide from the mass of the potassium iodide. Find the percent yield.

B. Percentage of Potassium Iodide in an Unknown Mixture

1. Obtain an unknown sample containing potassium iodide. Record the unknown number in the Data Table.

2. Repeat Steps 1–7 in the above procedure substituting the unknown mixture for potassium iodide.

3. Calculate the mass of potassium iodide in the unknown sample from the mass of precipitate. Find the percentage of potassium iodide in the unknown mixture.

PRELABORATORY ASSIGNMENT*

1. In your own words, define the following terms: actual yield, digestion, filtrate, percent yield, precipitate, stoichiometry, supernate, theoretical yield.

2. What problem arises if the precipitate is not allowed to settle completely from solution?

3. What is the purpose of the rubber policeman?

4. What should you do if particles of precipitate appear in the filtrate?

5. What are the primary sources of error in this experiment?

6. Is it possible to have a percent yield which is greater than 100%?

7. What safety precautions must be observed in this experiment?

*Answers in Appendix II.

DATA TABLE FOR ANALYSIS BY PRECIPITATION

A. Percent Yield of Lead(II) Iodide

mass of beaker + KI _____ g

mass of beaker _____ g

mass of KI _____ g

mass of filter paper + PbI_2 ppt _____ g

mass of filter paper _____ g

mass of PbI_2 ppt (actual yield) _____ g

Show the calculation for the theoretical yield of lead(II) iodide.

mass of PbI_2 ppt (theoretical yield) _____ g

Show the calculation for percent yield of lead(II) iodide.

Percent yield of PbI_2 _____ %

B. Percentage of Potassium Iodide in an Unknown Mixture

UNKNOWN # _____

mass of beaker + unknown mixture _____ g

mass of beaker _____ g

mass of unknown mixture _____ g

mass of filter paper + PbI_2 ppt _____ g

mass of filter paper _____ g

mass of PbI_2 ppt _____ g

Show the calculation for the mass of potassium iodide in the unknown mixture.

mass of KI _____ g

Show the calculation for percentage of potassium iodide in the unknown mixture.

Percentage of KI _____ %

1. A 1.020 g sample of sodium fluoride is dissolved in water and then precipitated with calcium nitrate solution. If the calcium fluoride precipitate weighs 0.905 g, what is the percent yield?

$$2\ NaF_{(aq)} + Ca(NO_3)_{2\,(aq)} \longrightarrow CaF_{2\,(s)} + 2\ NaNO_{3\,(aq)}$$

2. The stannous fluoride in a 10.000 g sample of toothpaste was extracted and then precipitated with lanthanum nitrate solution. If the mass of the precipitate is 0.105 g, what is the percentage of stannous fluoride in the toothpaste sample?

$$3\ SnF_{2\,(aq)} + 2\ La(NO_3)_{3\,(aq)} \longrightarrow 2\ LaF_{3\,(s)} + 3\ Sn(NO_3)_{2\,(aq)}$$

3. Limestone samples are mainly calcium carbonate. When a 0.750 g limestone is dissolved in hydrochloric acid, 165 mL of carbon dioxide gas is evolved at STP. What is the percentage of calcium carbonate in the limestone? The equation for the reaction is

$$CaCO_{3(s)} + 2\ HCl_{(aq)} \longrightarrow CaCl_{2(aq)} + H_2O_{(l)} + CO_{2(g)}$$

4. What is the minimum STP volume of hydrogen sulfide gas required to precipitate 0.555 g of bismuth sulfide? The equation for the reaction is

$$2\ Bi(NO_3)_{3(aq)} + 3\ H_2S_{(g)} \longrightarrow Bi_2S_{3(s)} + 6\ HNO_{3(aq)}$$

5. (optional) The insecticide lindane, $C_6H_6Cl_6$, is prepared from benzene and chlorine gas in the presence of ultraviolet light. If 150 mL of chlorine gas at STP is reacted with 150 mg of benzene, C_6H_6, what is the mass of lindane produced? The equation for the reaction is

$$C_6H_{6(l)} + 3\ Cl_{2(g)} \xrightarrow{\ UV\ } C_6H_6Cl_{6(s)}$$

Molar Volume of a Gas

OBJECTIVES

1. To determine experimentally the molar volume of hydrogen gas at standard temperature and pressure.
2. To determine the atomic mass of an unknown metal.
3. To be able to apply Dalton's Law of Partial Pressures, the combined gas laws, and mass-volume stoichiometry rules.
4. To develop the laboratory skills of collecting a gas over water and reading a barometer.

DISCUSSION

The *molar volume* of a gas is defined as the volume occupied by one mole of gas at standard temperature and pressure (STP). The theoretical value for the molar volume of an ideal gas at standard conditions is *22.4 liters per mole*. Since this is the volume occupied by *one mole*, it follows that the volume contains Avogadro's number of molecules. The mass of gas in the molar volume would be equal to the gram-molecular mass of the gas. A molar volume of hydrogen is illustrated in Figure 17-1.

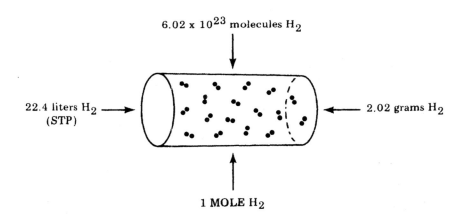

FIGURE 17-1　Mole concept applied to hydrogen gas.

A metal and an acid solution react to give a salt and hydrogen gas. In this experiment magnesium reacts with hydrochloric acid according to the equation

$$Mg_{(s)} + 2\,HCl_{(aq)} \longrightarrow MgCl_{2\,(aq)} + H_{2\,(g)}$$

An experimental value for the molar volume of hydrogen gas can be determined by considering the stoichiometry of the reaction.

PROBLEM EXAMPLE 17-1

A 0.0751 g sample of magnesium reacted with excess hydrochloric acid to produce 77.6 mL of hydrogen gas. The gas was collected over water at 21°C and an atmospheric pressure of 763 torr. Calculate the experimental molar volume of hydrogen gas at STP.

Solution: First, we must correct the experimental volume to STP. Since the hydrogen was collected over water, both hydrogen and water vapor contribute to the total gas pressure. We know the vapor pressure of water at 21°C is 18.6 torr (Table 17-1). The total pressure of the two gases is equal to the atmospheric pressure measured by the barometer. Applying Dalton's Law of Partial Pressures:

$$P_{atm} = P_{H_2} + P_{H_2O}$$
$$763\ torr = P_{H_2} + 18.6\ torr$$
$$P_{H_2} = 763\ torr - 18.6\ torr = 744\ torr$$

Before correcting the volume of hydrogen to STP, let us prepare a table for the data.

	V	P	T
old	77.6 mL	744 torr	21 + 273 = 294 K
new	V_{STP}	760 torr	0 + 273 = 273 K

We will apply the combined gas law in order to find the STP volume.

$$V_{STP} = V_{old} \times P_{factor} \times T_{factor}$$

P_{factor}: *the pressure increases so volume decreases*

T_{factor}: *the temperature decreases so volume decreases*

Since volume decreases for pressure and temperature changes in this example, both factors must be less than one.

$$V_{STP} = 77.6\ mL \times \frac{744\ torr}{760\ torr} \times \frac{273\ K}{294\ K} = 70.5\ mL$$

From the balanced equation we note that one mole of hydrogen is produced for every one mole of magnesium reacting; thus

$$0.0751\ g\ Mg \times \frac{1\ mole\ Mg}{24.3\ g\ Mg} \times \frac{1\ mole\ H_2}{1\ mole\ Mg} = 0.00309\ mole\ H_2$$

The molar volume is simply the ratio of the STP volume to the moles of hydrogen gas.

$$\frac{\text{liters of } H_2}{\text{moles of } H_2} = \text{molar volume of } H_2$$

$$\frac{70.5 \text{ mL } H_2}{0.00309 \text{ mole } H_2} \times \frac{1 \text{ L}}{1000 \text{ mL}} = 22.8 \text{ L/mole}$$

The experimental molar volume is 22.8 L/mole at STP. This compares favorably with the theoretical value of 22.4 L/mole.

After reacting magnesium metal and hydrochloric acid, we will react an unknown metal *(M)* in a similar fashion. The equation for the reaction is

$$M_{(s)} + 2 \text{ HCl}_{(aq)} \longrightarrow MCl_{2(aq)} + H_{2(g)}$$

By taking advantage of the molar volume concept, we can calculate the atomic mass of the unknown metal.

PROBLEM EXAMPLE 17-2

A 0.173 g sample of unknown metal reacted and evolved 65.5 mL of hydrogen gas. The gas was collected over water at 23°C and an atmospheric pressure of 766 torr. Calculate the atomic mass of the metal *(M)*.

Solution: The partial pressure of hydrogen is found by applying Dalton's Law. The vapor pressure of water at 23°C is 21.1 torr.

$$P_{H_2} = 766 \text{ torr} - 21.1 \text{ torr} = 745 \text{ torr}$$

To correct the volume of hydrogen to STP conditions, let's prepare a table for the data.

	V	P	T
old	65.5 mL	745 torr	23 + 273 = 296 K
new	V_{STP}	$76\bar{0}$ torr	0 + 273 = 273 K

Next, we will apply the combined gas law to find the STP volume.

$$V_{STP} = V_{old} \times P_{factor} \times T_{factor}$$

P_{factor}: *the pressure increases so volume decreases*

T_{factor}: *the temperature decreases so volume decreases*

In this example the volume decreases for both the pressure and temperature changes. Therefore, each factor is less than one.

$$V_{STP} = 65.5 \text{ mL} \times \frac{745 \text{ torr}}{76\bar{0} \text{ torr}} \times \frac{273 \text{ K}}{296 \text{ K}} = 59.2 \text{ mL}$$

According to the balanced equation for the reaction, one mole of unknown metal *(M)* produces one mole of hydrogen. The moles of unknown metal are determined accordingly.

$$59.2 \text{ mL H}_2 \text{ (STP)} \times \frac{1 \text{ L}}{1000 \text{ mL}} \times \frac{1 \text{ mole H}_2}{22.4 \text{ L H}_2} \times \frac{1 \text{ mole } M}{1 \text{ mole H}_2}$$

$$= 0.00264 \text{ mole } M$$

The gram-atomic mass of the unknown metal is expressed by the ratio of the mass of sample to the moles of metal.

$$\frac{\text{grams of } M}{\text{moles of } M} = \text{gram-atomic mass of } M$$

$$\frac{0.173 \text{ g } M}{0.00264 \text{ mole } M} = 65.5 \text{ g/ mole}$$

The atomic mass of the unknown metal *(M)* is 65.5 amu.

___

In this experiment we will collect hydrogen gas over water. The hydrogen displaces a volume of water that can be measured. Figure 17-2 illustrates an apparatus for collecting hydrogen.

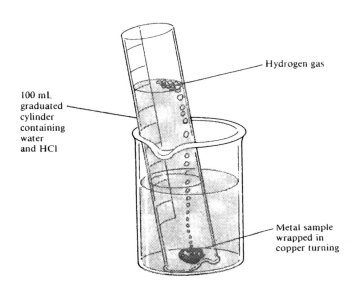

100 mL graduated cylinder containing water and HCl

Hydrogen gas

Metal sample wrapped in copper turning

FIGURE 17-2   Apparatus for collecting hydrogen gas over water.

Hydrochloric acid, HCl, is added to the graduated cylinder and then filled with water. The cylinder is inverted over a metal sample. As the acid diffuses to the bottom of the beaker, it begins to react with the metal sample. Bubbles of hydrogen gas are produced that displace water from the cylinder. The volume of gas can be read directly from the graduated cylinder.

If available, the hydrogen gas can be collected in a gas buret. The principle is identical to the graduated cylinder above. Figure 17-3 illustrates an alternate apparatus using a gas buret.

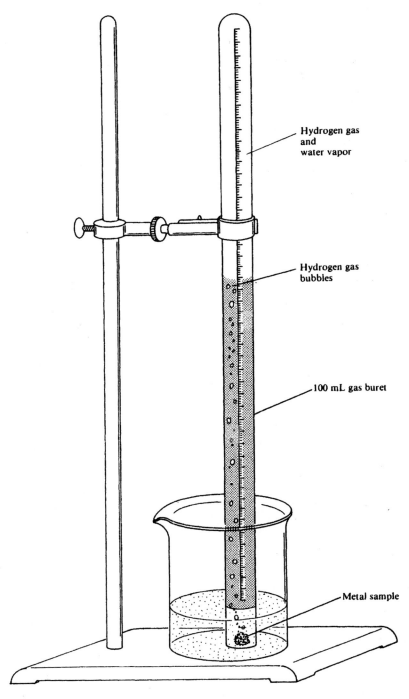

Hydrogen gas
and
water vapor

Hydrogen gas
bubbles

100 mL gas buret

Metal sample

FIGURE 17-3   Apparatus for collecting hydrogen over water in a gas buret.

EQUIPMENT AND CHEMICALS

Equipment

- 1000 mL beaker
- 100 mL graduated cylinder
- wash bottle
- 110°C thermometer
- barometer
- milligram balances
- tenth milligram balances (optional)
- 100 mL gas buret (optional)

Chemicals

- magnesium, Mg ribbon
- copper, Cu light turnings
- dilute hydrochloric acid, 6 M HCl
- unknown metal samples

PROCEDURE

A. Molar Volume of Hydrogen Gas

1. Cut a strip of magnesium ribbon having a mass between 0.07 and 0.09 g. (This corresponds to a 7-9 cm strip of magnesium ribbon). Weigh the metal accurately. Roll the ribbon into a compact coil and completely wrap the metal with strands of copper turnings.
2. Add 700 mL of water into a 1000 mL beaker. Drop the copper-wrapped metal into the water.
3. Add 25 mL of dilute hydrochloric acid into a 100 mL graduated cylinder. Fill to the uppermost rim with water, using a wash bottle. Place a small piece of paper towel over the entire rim and allow it to absorb water. Invert the graduated cylinder over the sink. Carefully put the graduated cylinder into the beaker. As the piece of towel floats free, place the cylinder over the copper-wrapped metal (Figure 17-2).

   NOTE: If the graduated cylinder loses water upon inversion, the piece of paper is too large. On occasion, the spout in the graduated cylinder is too deep. Exchanging the cylinder solves this problem.

   NOTE: If available, a gas buret can be substituted for the graduated cylinder.

4. Gas bubbles are observed when the acid diffuses down to the metal sample. After the reaction is complete, hold the cylinder vertically and read the bottom of the meniscus. Record the volume to the nearest 0.1 mL in the Data Table.
5. Place the thermometer into the water in the beaker. Record the temperature of the hydrogen gas in the Data Table.

   NOTE: The temperature of the water should agree with that of the gas.

6. Read the barometer and record the atmospheric pressure. Find the vapor pressure of water in Table 17-1.
7. Calculate the molar volume of hydrogen at STP.

B. Atomic Mass of an Unknown Metal

1. Obtain a sample of unknown metal *(M)* and record the number. The instructor will indicate the maximum sample mass for your unknown.
2. Follow the steps in the procedure above.

   NOTE: If the unknown metal is in very small pieces, it is convenient to stick the pieces onto tape and then wrap with copper strands. Do not seal the metal inside the tape.

NOTE: If the unknown metal is quite slow to react, store the apparatus in your laboratory locker until the next period.

3. Calculate the atomic mass of the unknown metal *(M)*.

PRELABORATORY ASSIGNMENT*

1. In your own words define the following terms: combined gas laws, Dalton's Law of Partial Pressures, molar volume, standard conditions, torr, vapor pressure.
2. Explain the phrase "collecting a gas over water."
3. Why must the mass of magnesium ribbon be no greater than 0.09 g?
4. How can you tell when the magnesium has reacted completely?
5. If the temperature of the water in the beaker is 23°C, what is the temperature of the hydrogen gas?
6. A sample of hydrogen gas is collected over water at 23°C and a barometer reading of 756 mm of mercury. What is the partial pressure of hydrogen in torr?
7. What are the major sources of error in this experiment?
8. What safety precautions should be observed in this experiment?

*Answers in Appendix II.

TABLE 17-1 VAPOR PRESSURE OF WATER

Temperature °C	Pressure torr (mm Hg)	Temperature °C	Pressure torr (mm Hg)
0	4.6	21	18.6
5	6.5	22	19.8
10	9.2	23	21.1
15	12.8	24	22.4
16	13.6	25	23.8
17	14.5	26	25.2
18	15.5	27	26.7
19	16.5	28	28.3
20	17.5	29	30.0
		30	31.8

DATA TABLE FOR MOLAR VOLUME OF A GAS

A. Molar Volume of Hydrogen Gas

mass of magnesium	_____ g	_____ g
volume of hydrogen	_____ mL	_____ mL
temperature of hydrogen	_____ °C	_____ °C
atmospheric pressure	_____ torr	_____ torr
vapor pressure of water	_____ torr	_____ torr
partial pressure of hydrogen	_____ torr	_____ torr

Correct the volume of hydrogen gas produced to STP conditions for trial 1.

volume of hydrogen (STP)	_____ mL	_____ mL

Show the calculation for the moles of hydrogen produced in trial 1.

moles of hydrogen	_____ mole	_____ mole

Show the molar volume calculation for trial 1.

Molar volume of hydrogen (STP)	_____ L/mole	_____ L/mole

B. Atomic Mass of an Unknown Metal UNKNOWN # _____

mass of metal *(M)* _____ g _____ g

volume of hydrogen _____ mL _____ mL

temperature of hydrogen _____ °C _____ °C

atmospheric pressure _____ torr _____ torr

vapor pressure of water _____ torr _____ torr

partial pressure of hydrogen _____ torr _____ torr

Correct the volume of hydrogen gas produced to STP conditions for trial 1.

volume of hydrogen (STP) _____ mL _____ mL

Show the calculation for the moles of unknown metal for trial 1.

moles of unknown metal _____ mole _____ mole

Show the calculation for the gram-atomic mass of the unknown metal for trial 1.

Gram-atomic mass of metal _____ g/mole _____ g/mole

Atomic mass of metal *(M)* _____ amu _____ amu

POSTLABORATORY ASSIGNMENT NAME ______________________

1. Mercuric oxide is decomposed according to the equation

$$2\ HgO_{(s)} \xrightarrow{\Delta} 2\ Hg_{(\ell)} + O_{2\,(g)}$$

A 2.154 g sample of HgO produces 125 mL of oxygen collected over water at 24° C and an atmospheric pressure of 763 torr. Calculate the STP molar volume for oxygen.

____

2. A 0.105 g sample of an unknown metal *(X)* reacted with hydrochloric acid according to the equation

$$2\ X_{(s)} + 6\ HCl_{(aq)} \longrightarrow 2\ XCl_{3\,(aq)} + 3\ H_{2\,(g)}$$

The volume of hydrogen collected over water was 144 mL at 22° C and a barometer reading of 764 torr. Calculate the atomic mass of the unknown metal.

____

Name the metal and explain your answer.

3. Why does the water remain in the inverted glass?

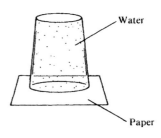

Water

Paper

4. A barometer reads 727 torr. Express the atmospheric pressure in:

(a) mm Hg

(b) cm Hg

(c) atm

(d) in. Hg

5. (optional) A barometer in Denver, Colorado reads 630 torr. What is the height in feet of a barometer containing water? (Hint: Mercury is 13.6 times as dense as water.)

Molecular Mass of a Gas

OBJECTIVES

1. To determine experimentally the molecular mass of nitrogen.
2. To determine the molecular mass of an unknown gas.
3. To determine the density of natural gas at standard conditions.
4. To develop the lab skills of transferring and weighing gases, and reading the barometer.

DISCUSSION

The principal laboratory skill in this experiment is finding the mass of a gas. A gas is difficult to handle as it diffuses rapidly and has a low density. To obtain the mass of a gas we have to place it in a container, such as a stoppered flask. This presents a problem because we must know the mass of the flask in order to weigh by difference. Ordinarily, we would weigh the flask empty, but the empty flask contains air. Furthermore, the mass of air is substantial when compared to the mass of another gas.

To solve this problem, we will calculate the mass of air in the flask. By subtracting this mass of air, we can find the mass of an empty flask.

PROBLEM EXAMPLE 18-1

A flask filled with air has a mass of 122.681 g. The volume of the flask is found to be 264 mL. The temperature is 23°C and the barometer reading is 77.0 cm of mercury. Calculate the mass of the empty flask.

Solution: Using the *Handbook of Chemistry and Physics*, we find the density of dry air at 23°C and 77.0 cm Hg is 0.001208 g/mL. The mass of air is calculated as follows:

$$264 \text{ mL air} \times \frac{0.001208 \text{ g air}}{1 \text{ mL air}} = 0.319 \text{ g air}$$

The mass of the empty flask is 122.681 g − 0.319 g = 122.362 g.

It is our objective to determine the molecular mass of nitrogen as well as an unknown gas. Let's suppose we fill the flask from the preceding example with nitrous oxide, that is, laughing gas. After collecting data, we can find the molecular mass of nitrous oxide.

PROBLEM EXAMPLE 18-2

Calculate the molecular mass of nitrous oxide if 264 mL of the gas measured at $23°C$ and $77\bar{0}$ torr has a mass of 0.491 g.

Solution: First, we will prepare a table of data and correct the volume to STP conditions.

	V	P	T
old	264 ml	$77\bar{0}$ torr	23 + 273 = 296 K
new	V_{STP}	$76\bar{0}$ torr	0 + 273 = 273 K

$$V_{STP} = V_{old} \times P_{factor} \times T_{factor}$$

P_{factor}: *the pressure decreases so volume increases*

T_{factor}: *the temperature decreases so volume decreases*

In this example the pressure and temperature changes result in a volume increase and decrease, respectively. Therefore, the pressure factor is greater than one and the temperature factor less than one.

$$V_{STP} = 264 \text{ mL} \times \frac{77\bar{0} \text{ torr}}{76\bar{0} \text{ torr}} \times \frac{273 \text{ K}}{296 \text{ K}} = 247 \text{ mL}$$

Second, we can calculate molecular mass of nitrous oxide:

$$\frac{0.491 \text{ g}}{247 \text{ mL (STP)}} \times \frac{1000 \text{ mL}}{1 \text{ L}} \times \frac{22.4 \text{ L (STP)}}{1 \text{ mole}} = 44.5 \text{ g /mole}$$

The experimental molecular mass for nitrous oxide is 44.5 amu. This compares favorably with the theoretical value of 44.0 amu for nitrous oxide, N_2O.

We will conclude the experiment by finding the density of natural gas in the laboratory. Natural gas is not a single compound but rather a mixture of several gases. Its composition varies but it is usually about 90% methane (CH_4) with lesser amounts of ethane (C_2H_6), propane (C_3H_8), and butane (C_4H_{10}). A trace of a foul-smelling sulfur compound is added for safety, that is, detection.

PROBLEM EXAMPLE 18-3

Calculate the density of natural gas at STP if 0.191 g has a volume of 264 mL at $20°C$ and 758 torr.

Solution: Let's set up a table for the data and then correct the volume to standard conditions.

	V	P	T
old	264 mL	758 torr	20 + 273 = 293 K
new	V_{STP}	760 torr	0 + 273 = 273 K

$$V_{STP} = V_{old} \times P_{factor} \times T_{factor}$$

P_{factor}: *the pressure increases so volume decreases*

T_{factor}: *the temperature decreases so volume decreases*

The volume decreases for the above changes in pressure and temperature, so both factors are less than one.

$$V_{STP} = 264 \text{ mL} \times \frac{758 \text{ torr}}{760 \text{ torr}} \times \frac{273 \text{ K}}{293 \text{ K}} = 245 \text{ mL}$$

Second, we can calculate the density of natural gas at STP:

$$\frac{0.191 \text{ g gas}}{245 \text{ mL}} \times \frac{1000 \text{ mL}}{1 \text{ L}} = 0.780 \text{ g/L}$$

EQUIPMENT AND CHEMICALS

Equipment

- 250 mL Erlenmeyer flasks (3) with rubber stoppers to fit
- 110°C thermometer
- barometer
- *Handbook of Chemistry and Physics*

Chemicals

- nitrogen gas cylinder
- unknown gas cylinder(s)
- natural gas

PROCEDURE

A. Collecting Experimental Data

1. Accurately weigh a 250 mL Erlenmeyer flask fitted with a rubber stopper.
2. Slowly flush the air out of the flask using nitrogen gas (Figure 18-1). This procedure requires 5 to 10 seconds. Remove the nitrogen tube and restopper the flask immediately. Weigh the flask, stopper, and nitrogen gas.

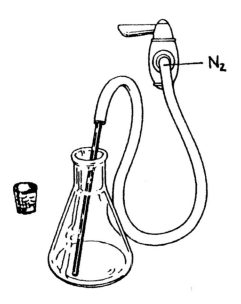

FIGURE 18-1   Replacing the air in a flask with nitrogen gas.

3. Slowly flush the nitrogen from the flask for 5 to 10 seconds using an unknown gas. Remove the unknown gas tube and re-stopper the flask immediately. Weigh the flask, stopper, and unknown gas.
4. Flush the unknown gas with natural gas. Re-stopper and weigh the flask, stopper, and natural gas.

> NOTE: For best results, hold the flask upside down and flush with natural gas. The density of natural gas is lighter than air.

5. Fill the flask with water and insert the rubber stopper slowly. Wipe off any water on the outside of the flask with a paper towel. Weigh the flask, stopper, and water to the nearest gram.

> NOTE: Since the density of water is 1.00 g/mL the mass of water is numerically equal to the volume of water in the flask. The volume of water is the same as the volume of each gas (air, nitrogen, unknown gas and natural gas).

6. Repeat the foregoing procedure with two additional 250 mL flasks.
7. Read the thermometer and record the temperature. Assume the temperature of the room and each gas are the same.
8. Read the barometer and record the pressure. Assume the pressure of the atmosphere and each gas are the same.
9. Refer to the *Handbook of Chemistry and Physics, Density of Dry Air*. Record the density of air for the experimental temperature and pressure.

> NOTE: The reference book lists pressure in centimeters of mercury. Convert atmospheric pressure from units of torr to cm Hg.

B. Calculating the Mass of an Empty Flask

1. Record the relevant information from Procedure A in the Data Table provided.
2. Calculate the mass of air in each flask. Find the mass of each empty flask by subtraction.

C. Molecular Mass of Nitrogen

1. Record the relevant information from Procedures A and B in the Data Table provided.
2. Correct the volume of nitrogen to STP and calculate the molecular mass.

D. Molecular Mass of an Unknown Gas

1. Record the relevant information from Procedures A and B in the Data Table provided.
2. Correct the volume of unknown gas to STP and calculate the molecular mass.

E. Density of Natural Gas at STP

1. Record the relevant information from Procedures A and B in the Data Table provided.
2. Correct the volume of natural gas to STP and calculate the density.

PRELABORATORY ASSIGNMENT*

1. In your own words define the following terms: atmospheric pressure, combined gas laws, experimental conditions, molar volume, standard conditions, torr, weighing by difference.
2. When introducing a gas into a flask, how can you tell if the previous gas is flushed out?
3. Why should some gases be introduced into an upside-down flask?
4. A barometer reading of 758 torr corresponds to what value in cm of Hg?
5. A flask containing air has a mass of 122.222 g and a mass of 387 g filled with water. What is the volume of air in the flask?
6. What are the main sources of error in this experiment?
7. What safety precautions should be observed in this experiment?

*Answers in Appendix II.

DATA TABLE FOR MOLECULAR MASS OF A GAS

A. Collecting Experimental Data

mass of flask and stopper + air	_____ g	_____ g	_____ g
mass of flask and stopper + nitrogen	_____ g	_____ g	_____ g
mass of flask and stopper + unknown gas	_____ g	_____ g	_____ g
mass of flask and stopper + natural gas	_____ g	_____ g	_____ g
mass of flask and stopper + water	_____ g	_____ g	_____ g
mass of water	_____ g	_____ g	_____ g
volume of water	_____ mL	_____ mL	_____ mL
volume of gas (air, nitrogen, etc)	_____ mL	_____ mL	_____ mL
temperature of gas	_____ °C		
pressure of gas	_____ torr		
density of air (see *Handbook*)	_____ g/mL		

B. Calculating the Mass of an Empty Flask

mass of flask and stopper + air (A)	_____ g	_____ g	_____ g

Using the volume and density of air in the preceding table, show the calculation for the mass of air in flask 1.

mass of air	_____ g	_____ g	_____ g
mass of flask and stopper (without air)	_____ g	_____ g	_____ g

C. Molecular Mass of Nitrogen

mass of flask and stopper + nitrogen (A)   _____ g   _____ g   _____ g

mass of flask and stopper (B)   _____ g   _____ g   _____ g

mass of nitrogen   _____ g   _____ g   _____ g

volume of nitrogen (A)   _____ mL   _____ mL   _____ mL

Correct the volume of nitrogen to STP for trial 1.

volume of nitrogen (STP)   _____ mL   _____ mL   _____ mL

Show the calculation for the gram-molecular mass of nitrogen for trial 1.

gram-molecular mass   _____ g/mole   _____ g/mole   _____ g/mole

molecular mass of nitrogen   _____ amu   _____ amu   _____ amu

D. Molecular Mass of an Unknown Gas   UNKNOWN # _____

mass of flask and stopper +
   unknown gas (A)   _____ g   _____ g   _____ g

mass of flask and stopper (B)   _____ g   _____ g   _____ g

mass of unknown gas   _____ g   _____ g   _____ g

volume of unknown gas (A)   _____ mL   _____ mL   _____ mL

Correct the volume of unknown gas to STP for trial 2.

volume of unknown gas (STP)   _____ mL   _____ mL   _____ mL

Show the calculation for the gram-molecular mass of the unknown gas for trial 2.

gram-molecular mass   _____ g/mole   _____ g/mole   _____ g/mole

molecular mass
   of unknown gas   _____ amu   _____ amu   _____ amu

E. Density of Natural Gas at STP

mass of flask and stopper
and natural gas (A) _____ g _____ g _____ g

mass of flask and stopper (B) _____ g _____ g _____ g

mass of natural gas _____ g _____ g _____ g

volume of natural gas (A) _____ mL _____ mL _____ mL

Correct the volume of natural gas to STP for trial 3.

volume of natural gas (STP) _____ mL _____ mL _____ mL

Show the calculation for the density of natural gas at STP for trial 3.

density of natural gas (STP) _____ g/L _____ g/L _____ g/L

POSTLABORATORY ASSIGNMENT   NAME _____

1.   A flask contains 246 mL of natural gas at STP. How many molecules are in the flask?

_____

2.   What is the volume at STP for 0.415 g of argon gas?

_____

3.   The density of ozone at STP is 2.14 g/L. Ozone contains only oxygen; find the molecular mass and formula of ozone.

_____

4.   A flask weighing 104.223 g contains 265 mL of moist air at 25°C and 760 torr. The density of moist air under these conditions is found in the *Handbook of Chemistry and Physics:* 1.1845 g/L. What is the mass of the flask without air?

_____

5. A 125 mL flask contains 0.239 g of acetone vapor at 766 torr and 100°C. Calculate the molecular mass of acetone.

_____

6. Methyl mercaptan is a foul-smelling gaseous compound added to natural gas in order to detect leaks. Calculate the STP density (g/L) if 1.037 g of the gas occupy a volume of 528 mL at 22°C and 754 torr.

_____

7. (optional) A 250 mL flask is found to contain 6.45×10^{21} molecules of nitrogen gas. The flask is then flushed with an unknown gas. How many molecules of unknown gas are in the flask if temperature and pressure remain constant?

Solutions

OBJECTIVES

1. To observe the solubility of ionic and covalent solutes.
2. To observe the miscibility of water and various solvents.
3. To study the factors that affect the rate of solubility.
4. To demonstrate the behavior of a supersaturated solution.
5. To determine the percent by mass and molar concentration of an unknown solution.
6. To become proficient in the laboratory skills of pipetting and evaporating a solution to dryness.

DISCUSSION

The phrase *'like dissolves like'* describes the general principle of solubility. That is, ionic and polar solutes dissolve in polar solvents. Nonpolar solutes dissolve in nonpolar solvents.

Water is a polar solvent. It dissolves ionic compounds such as table salt, NaCl, and polar compounds such as sugar, $C_{12}H_{22}O_{11}$. Carbon tetrachloride is a nonpolar solvent and dissolves nonpolar compounds. Carbon tetrachloride is therefore not a solvent for salt or sugar. Nonpolar solvents do not dissolve ionic or polar compounds.

Liquids that dissolve in one another are said to be *miscible*. Once again, the general principle of *'like dissolves like'* dictates whether two liquids are miscible. If both liquids have polar bonds, they are miscible. Two liquids that are nonpolar also dissolve in one another. However, a polar liquid and a nonpolar liquid are immiscible. They repel one another and separate into two layers.

The process of dissolving is simply the action of the solvent on the individual solute particles. The solvent molecules attract the particles of solute, and draw them into solution. The *rate of solubility* is obviously dependent upon the rate at which the solvent attacks the solute. In the second part of this experiment, we will study how the three factors—particle size, stirring, and temperature—affect the rate of solubility.

The amount of solute dissolved in a given amount of solution is termed concentration. Concentration can be *qualitatively* expressed using the terms dilute and concentrated. Saturated, unsaturated, and supersaturated are also used to indicate concentration. A *saturated* solution is one in which the maximum amount of solute is dissolved in solution for a given temperature. *Unsaturated* denotes less than maximum solute concentration. *Supersaturated* refers to the condition in which an excess amount of solute is dissolved for a given temperature.

Percent by mass and molarity are two ways of expressing the concentration of a solution quantitatively. The *percent by mass* concentration expresses the ratio of the mass of solute to the mass of solution.

$$\% \text{ by mass} = \frac{\text{mass of solute}}{\text{mass of solution}} \times 100$$

The molar concentration is defined as the number of moles of solute per liter of solution. The molar concentration is also referred to as *molarity* (M), which we can express in equation form.

$$M = \frac{\text{moles of solute}}{\text{liter of solution}}$$

PROBLEM EXAMPLE 19-1

A 10.0 mL sample of sodium chloride solution has a mass of 10.214 g. After evaporating to dryness, the solute residue weighed 0.305 g. Calculate the percent by mass concentration of the solution.

Solution: The percent by mass concentration is readily obtained using the above equation.

$$\frac{0.305 \text{ g NaCl}}{10.214 \text{ g solution}} \times 100 = 2.99\% \text{ NaCl}$$

PROBLEM EXAMPLE 19-2

Calculate the molar concentration of the sodium chloride solution in Problem Example 19-1.

Solution: The molecular mass of NaCl is 58.5 amu; therefore, the molarity is

$$\frac{0.305 \text{ g NaCl}}{10.0 \text{ mL solution}} \times \frac{1 \text{ mole NaCl}}{58.5 \text{ g NaCl}} \times \frac{1000 \text{ mL}}{1 \text{ L}} = \frac{0.521 \text{ mole NaCl}}{1 \text{ L solution}} = 0.521 \text{ M NaCl}$$

EQUIPMENT AND CHEMICALS

Equipment

- 16 × 150 mm test tubes (6)
- test tube rack
- glass stirring rod
- mortar and pestle
- wire gauze
- wash bottle
- evaporating dish
- 250 mL beaker
- 10 mL pipet
- pipet bulb
- 100 mL beaker

Chemicals

- potassium permanganate, $KMnO_4$
- iodine crystals, I_2
- sucrose crystals, $C_{12}H_{22}O_{11}$
- hexane, C_6H_{14}
- methanol, CH_3OH
- acetone, C_3H_6O
- heptane, C_7H_{16}
- ethanol, C_2H_5OH
- rock salt, NaCl
- sodium acetate trihydrate, $NaC_2H_3O_2 \cdot 3H_2O$
- unknown sodium chloride solutions (3-5% NaCl)

PROCEDURE

A. Solutes and Solvents

1. *Solubility.* Place six *dry* test tubes in a test tube rack. Drop a small crystal of potassium permanganate, $KMnO_4$, into each of three test tubes. Add 2 mL (1/10 test tube) of water into the first, hexane into the second, and methanol into the third. Record whether the crystal is soluble (*s*) or insoluble (*ins*) in each solvent.

   Repeat the above procedure in the other three *dry* test tubes substituting iodine, I_2, as the crystal.

2. *Miscibility.* Put 2 mL of water into each of three test tubes. Add 2 mL of acetone into the first, heptane into the second, and ethanol into the third. Shake vigorously and record whether water is miscible (*misc*) or immiscible (*immisc*) in each solvent.

B. Rate of Dissolving

1. Set up a waterbath with distilled water as shown in Figure 19-1. Half fill three test tubes with water. Put the test tubes in the beaker and bring to a boil. After boiling, shut off the heat.

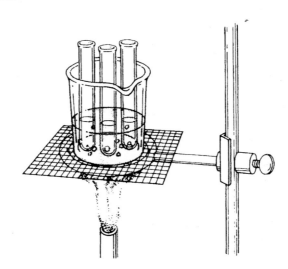

FIGURE 19-1 Waterbath apparatus for heating solutions in three test tubes.

2. Half fill a fourth test tube with distilled water. Add a single crystal of rock salt and record the time for dissolving in the Data Table.
3. Add a crystal of rock salt into a test tube in the waterbath. Record the length of time for the crystal to dissolve.
4. Add a crystal of rock salt into a test tube in the waterbath. Stir the solution continuously and record the dissolving time.
5. Grind a crystal of rock salt with the mortar and pestle. Add the powder into the remaining test tube in the waterbath. Stir the solution and record the dissolving time.

C. Demonstration of Supersaturation

1. Add crystals of sodium acetate trihydrate, $NaC_2H_3O_2 \cdot 3H_2O$, into a test tube until it is 1/4 full. Add just enough distilled water to cover the crystals.
2. Set up the apparatus shown in Figure 19-2. Bring the water to a boil and then remove the heat. Put the test tube into the waterbath and stir the solution with a glass rod.

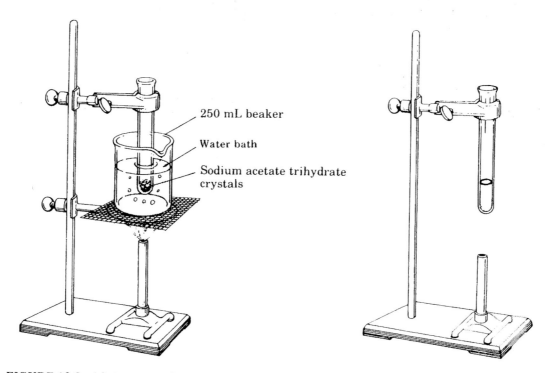

FIGURE 19-2 **(a)** Apparatus for dissolving the sodium acetate trihydrate crystals; **(b)** and allowing the supersaturated solution to cool.

When the crystals are dissolved, move up the utility clamp so the test tube is out of the beaker.

3. After the test tube feels cool, drop a tiny seed crystal of sodium acetate into the solution and observe the results.

D. Concentration of a Sodium Chloride Solution

1. Obtain about 25 mL of unknown sodium chloride solution in a dry 100 mL beaker and record the unknown number.
2. Weigh a dry evaporating dish on the balance. 46.47
3. Condition a pipet and transfer a 10.0 mL sample of unknown solution into the evaporating dish. Reweigh the dish and solution. 56.09
4. Add about 200 mL of distilled water into a 250 mL beaker. Place the evaporating dish in the beaker and evaporate the solution to dryness (see Figure 19-3).
5. After evaporation, remove the dish and wipe the bottom of the dish dry. Hold the dish with crucible tongs over a flame to dry the last traces of moisture. Allow the dish to cool and weigh the evaporating dish with the solute residue.

   NOTE: Do not heat the dish too strongly as this may cause some of the residue to pop from the dish.

6. Calculate the percent by mass and molar concentration of the unknown sodium chloride solution.

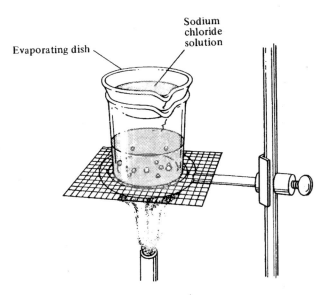

FIGURE 19-3   Apparatus for evaporating the sodium chloride
solution to dryness.

PRELABORATORY ASSIGNMENT*

1. In your own words define the following terms: conditioning, immiscible, miscible, molarity, nonpolar compound, percent by mass concentration, polar compound, saturated solution, seed crystal, solute, solvent, solution, supersaturated solution, unsaturated solution.
2. Explain the 'like dissolves like' rule.
3. How can you tell if a crystal is slightly soluble in a solvent?
4. Why is distilled water used in the waterbath?
5. Why is it necessary to stir the supersaturated solution after heating and before cooling? Why must you avoid jarring the test tube as it cools?
6. A 10.0 mL sample of unknown sodium chloride solution was analyzed and gave the following data:

> mass of evaporating dish + solution   50.827 g
> mass of evaporating dish   40.505 g
> mass of evaporating dish + solute   40.969 g

Calculate the (a) percent by mass and (b) molar concentration of the solution.
7. In determining the concentration of the sodium chloride solution, what are the major sources of error?
8. What safety precautions should be observed in this experiment?

*Answers in Appendix II.

Neutralization Titration

OBJECTIVES

1. To standardize a sodium hydroxide solution with potassium hydrogen phthalate.
2. To determine the molar and percent by mass concentration of acetic acid in an unknown vinegar solution.
3. To gain experience in applying solution stoichiometry rules to an acid-base titration.
4. To acquire proficiency in the laboratory techniques of pipeting, titrating, and using phenolphthalein indicator.

DISCUSSION

A *titration* measures the volume of solution delivered from a buret. In this experiment sodium hydroxide is titrated into a flask containing an acid. After a sufficient amount of base is added to neutralize the acid in the flask, we will stop the titration. This is termed the *endpoint* and is signaled by an *indicator* that changes color. The indicator used in this experiment is phenolphthalein. Phenolphthalein is colorless in acid and red-pink in base. Therefore, the acid solution containing the indicator will be colorless until a very slight excess of base is titrated. At the endpoint the phenolphthalein changes the color of the solution to red-pink. A single drop of base is sufficient to bring about the color change. Figure 20-1 illustrates the sequence of steps in the titration.

This experiment begins with the dilution of 6 M sodium hydroxide to a concentration of about 0.3 M. Since the dilution provides only an approximate concentration, it is necessary to *standardize* the solution. That is, we wish to determine the sodium hydroxide concentration to three significant digits. We will choose solid crystals of potassium hydrogen phthalate (100 percent purity) for our standard acid.

The formula for the acid is $KHC_8H_4O_4$ although it is usually abbreviated KHP. After dissolving the acid in water, the base is standardized according to the equation:

$$KHP_{(aq)} + NaOH_{(aq)} \longrightarrow KNaP_{(aq)} + H_2O_{(\ell)}$$

PROBLEM EXAMPLE 20-1

1.555 g of pure KHP (mm = 204 amu) are dissolved in water and the solution is titrated with 29.60 mL of sodium hydroxide to a phenolphthalein endpoint. Find the molarity of the sodium hydroxide solution.

Solution: Referring to the preceding equation for the reaction and applying the rules of stoichiometry,

$$1.555 \text{ g KHP} \times \frac{1 \text{ mole KHP}}{204 \text{ g KHP}} \times \frac{1 \text{ mole NaOH}}{1 \text{ mole KHP}} = 0.00762 \text{ mole NaOH}$$

The molarity of the NaOH is found by

$$\frac{0.00762 \text{ mole NaOH}}{29.60 \text{ mL solution}} \times \frac{1000 \text{ mL}}{1 \text{ L}} = \frac{0.257 \text{ mole NaOH}}{1 \text{ L solution}} = 0.257 \text{ M NaOH}$$

The concentration of the standardized NaOH is 0.257 M. This agrees reasonably with the approximate concentration of 0.3 M from dilution.

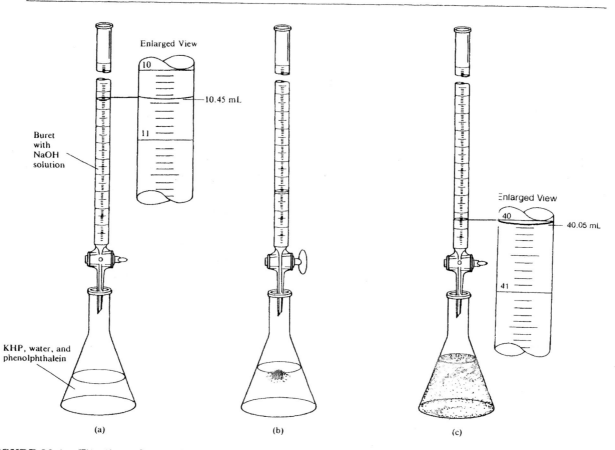

FIGURE 20-1 Titration of potassium hydrogen phthalate with sodium hydroxide solution. (a) Read the initial volume of base in the buret (10.45 mL). (b) Temporary flashes of pink indicate the endpoint is approaching. (c) A permanent pink color signals the base has neutralized the acid. Read the final volume of base in the buret (40.05 mL). The volume of NaOH required for the KHP sample is 40.05 mL − 10.45 mL = 29.60 mL.

After standardizing the sodium hydroxide solution, we will determine the acetic acid concentration in an unknown vinegar solution. A sample of vinegar will be titrated with the standard sodium hydroxide to a phenolphthalein endpoint. The equation for the reaction is

$$HC_2H_3O_2{}_{(aq)} + NaOH_{(aq)} \longrightarrow NaC_2H_3O_2{}_{(aq)} + H_2O_{(\ell)}$$

PROBLEM EXAMPLE 20-2

In the titration of a 10.0 mL vinegar sample, 35.05 mL of the above standard 0.257 M sodium hydroxide was required. Calculate the (a) molar and (b) percent by mass concentration of the acetic acid.

Solution:

(a) The moles of acetic acid titrated is

$$35.05 \text{ mL solution} \times \frac{1 \text{ L}}{1000 \text{ mL}} \times \frac{0.257 \text{ mole NaOH}}{1 \text{ L solution}} \times \frac{1 \text{ mole HC}_2\text{H}_3\text{O}_2}{1 \text{ mole NaOH}}$$

$$= 0.00901 \text{ mole HC}_2\text{H}_3\text{O}_2$$

and the molarity of $HC_2H_3O_2$ is

$$\frac{0.00901 \text{ mole HC}_2\text{H}_3\text{O}_2}{10.0 \text{ mL solution}} \times \frac{1000 \text{ mL}}{1 \text{ L}} = \frac{0.901 \text{ mole HC}_2\text{H}_3\text{O}_2}{1 \text{ L solution}} = 0.901 \text{ M}$$

(b) To calculate the percent by mass concentration, we must know the density of the vinegar solution, 1.01 g/mL, and the molecular mass of acetic acid, 60.0 amu.

$$\frac{0.901 \text{ mole HC}_2\text{H}_3\text{O}_2}{1 \text{ L solution}} \times \frac{60.0 \text{ g HC}_2\text{H}_3\text{O}_2}{1 \text{ mole HC}_2\text{H}_3\text{O}_2} \times \frac{1 \text{ L}}{1000 \text{ mL}} \times \frac{1 \text{ mL solution}}{1.01 \text{ g solution}} \times 100$$

$$= 5.35\% \text{ HC}_2\text{H}_3\text{O}_2$$

EQUIPMENT AND CHEMICALS

Equipment

- graduated cylinder
- 1000 mL Florence flask with stopper to fit
- 125 mL Erlenmeyer flasks (3)
- funnel
- buret stand (or ring stand)
- buret clamp (or utility clamp)
- 50 mL buret
- 10 mL pipet
- pipet bulb
- 100 mL beaker
- wash bottle with distilled water

Chemicals

- dilute sodium hydroxide, 6 M NaOH
- potassium hydrogen phthalate, solid $KHC_8H_4O_4$ (KHP)
- phenolphthalein indicator
- unknown vinegar solutions (4-6% $HC_2H_3O_2$)

PROCEDURE

A. Preparation of Standard Sodium Hydroxide Solution

1. Measure 25 mL of dilute sodium hydroxide (6 M NaOH) into a graduated cylinder and transfer to a 1000 mL Florence flask half filled with distilled water. Stopper the flask and swirl until homogeneous.

2. Condition a buret with the sodium hydroxide solution in the Florence flask. Place the buret in a clamp on a stand. Close the stopcock. Add sodium hydroxide solution into the buret through a small funnel.

   NOTE: Care should be taken not to overfill the buret.

3. Number three 125 mL Erlenmeyer flasks for identification. Record the mass of each flask. Accurately weigh out 1.2-1.8 g samples of KHP. Add about 25 mL of distilled water to each sample and heat as necessary to dissolve the acid crystals.

4. Prepare for the titration by following these steps.

- *Clear the buret tip of bubbles.*
- *Place one of the 125 mL flasks under the buret and position the tip as shown in Figure 20-1.*
- *Observe the meniscus and record the initial buret reading.*
- *Add two drops of phenolphthalein to each flask.*

Titrate the KHP sample until a permanent red-pink endpoint is reached. Record the final buret reading.
5. Refill the buret, record the initial volume reading, and titrate the second KHP sample.
6. Repeat the titration for the third KHP sample.

NOTE: SAVE THE SODIUM HYDROXIDE SOLUTION FOR PROCEDURE B.

7. Calculate the molarity of the sodium hydroxide solution for each trial. Record the average molarity in the Data Table of Procedure B.

B. Concentration of Acetic Acid in a Vinegar Solution

1. Obtain about 50 mL of unknown vinegar solution in a dry 100 mL beaker and record the unknown number.
2. Condition a pipet. Transfer a 10.0 mL sample of unknown solution into each of three 125 mL flasks.

NOTE: It is not necessary to dry the flasks since they will not be weighed.

Add 25 mL of distilled water and two drops of phenolphthalein into each flask.
3. Refill the buret and record the initial volume. Titrate the acetic acid in the first sample to a red-pink endpoint. Record the final buret reading.
4. Refill the buret, record the initial volume reading, and titrate the second vinegar sample.
5. Repeat the titration for the third sample.
6. Calculate the molarity of acetic acid in the unknown vinegar solution.
7. Convert the molar concentration to percent by mass. Assume a density of 1.01 g/mL for the unknown solution.

NOTE: When the titrations are completed, rinse the buret with several portions of distilled water to remove all traces of the caustic sodium hydroxide solution.

PRELABORATORY ASSIGNMENT*

1. In your own words define the following terms: conditioning, endpoint, indicator, molarity, percent by mass concentration, standardization, titration.

2. Which of the following is a serious source of experimental error?
   - (a) The sodium hydroxide is not mixed thoroughly in the Florence flask.
   - (b) The Florence flask is left unstoppered.
   - (c) The buret is not conditioned.
   - (d) The KHP samples are dissolved in 35 mL (not 25 mL) of distilled water.
   - (e) Three (not two) drops of phenolphthalein indicator are used.
   - (f) Bubbles are not cleared from the tip of the buret.
   - (g) The Erlenmeyer flasks are not dried before weighing the KHP samples.
   - (h) The Erlenmeyer flasks are not dried before pipetting the vinegar samples.

3. Observe and record the following buret readings.

(a) 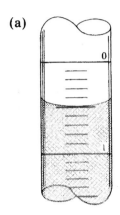 (b)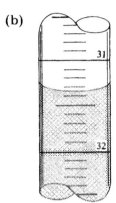

4. How can you tell when the endpoint is near? How much base is required at the end-point to flip the indicator from colorless to pink?

5. If the first KHP sample required 27.30 mL to reach an endpoint, what volume should be required for the second and third samples?

6. If the first 10.0 mL vinegar sample required 30.15 mL to reach an endpoint, what volume should be required for the second and third samples?

7. What safety precautions should be observed in this experiment?

*Answers in Appendix II.

DATA TABLE FOR NEUTRALIZATION TITRATION

A. Preparation of Standard Sodium Hydroxide Solution

mass of Erlenmeyer flask + KHP	_____ g	_____ g	_____ g
mass of Erlenmeyer flask	_____ g	_____ g	_____ g
mass of KHP	_____ g	_____ g	_____ g
final buret reading	_____ mL	_____ mL	_____ mL
initial buret reading	_____ mL	_____ mL	_____ mL
volume NaOH	_____ mL	_____ mL	_____ mL

Show the calculation for the molarity of sodium hydroxide for trial 1.

Molarity of NaOH	_____ M	_____ M	_____ M
Average molarity of NaOH		_____ M	

B. **Concentration of Acetic Acid in a Vinegar Solution**   UNKNOWN # _____

Average molarity of NaOH (see Procedure A)   _____ M

volume $HC_2H_3O_2$ sample	_____ mL		_____ mL		_____ mL
final buret reading	_____ mL		_____ mL		_____ mL
initial buret reading	_____ mL		_____ mL		_____ mL
volume NaOH	_____ mL		_____ mL		_____ mL

Show the calculation for the molarity of acetic acid for sample 1.

Molarity of $HC_2H_3O_2$	_____ M		_____ M		_____ M

Show the calculation for the percent by mass concentration of acetic acid for sample 1.

Percent by mass concentration of $HC_2H_3O_2$	_____ %		_____ %		_____ %

Average percent by mass $HC_2H_3O_2$ _____ %

POSTLABORATORY ASSIGNMENT NAME ______________________

1. A hydrochloric acid solution is standardized using 0.502 g of sodium carbonate. Find the molarity of the acid if 30.50 mL are required for the titration.

$$2\ HCl_{(aq)} + Na_2CO_{3(aq)} \longrightarrow 2\ NaCl_{(aq)} + H_2O_{(\ell)} + CO_{2(g)}$$

____

2. A 10.0 mL sample of household ammonia solution required 38.50 mL of 0.311 M HCl to achieve neutralization. Calculate (a) the molar concentration of the ammonia solution and (b) convert to percent by mass concentration of ammonia (17.0 amu), given a solution density of 0.983 g/mL.

$$HCl_{(aq)} + NH_3 \cdot H_2O_{(aq)} \longrightarrow NH_4Cl_{(aq)} + H_2O_{(\ell)}$$

(a) ______________________

(b) ______________________

3. If 19.65 mL of 0.145 M nitric acid is required to neutralize 50.0 mL of barium hydroxide solution, what is the molar concentration of the base?

$$2\ HNO_{3(aq)} + Ba(OH)_{2\,(aq)} \longrightarrow Ba(NO_3)_{2(aq)} + 2\ H_2O_{(\ell)}$$

____

4. A Rolaids tablet contains calcium carbonate to neutralize stomach acid. If 44.55 mL of 0.448 M hydrochloric acid is required to neutralize one tablet, how many milligrams of calcium carbonate are in a Rolaids tablet?

$$CaCO_{3(s)} + 2\ HCl_{(aq)} \longrightarrow CaCl_{2(aq)} + H_2O_{(\ell)} + CO_{2(g)}$$

5. (optional) A student carefully diluted 25.0 mL of 6 M NaOH solution in 475 mL of distilled water. Calculate the molarity of the diluted solution of base.

Explain why this diluted NaOH solution cannot be used as a standard solution of base.

Ionic Equations

OBJECTIVES

1. To gain experience in observing the electrical conductivity of ionic and molecular substances.
2. To determine if a substance is a strong electrolyte, weak electrolyte, or nonelectrolyte.
3. To follow the course of a chemical reaction by observing electrical conductivity.
4. To become proficient in writing net ionic equations.

DISCUSSION

Electrical conductivity is based on the principle of electron movement from one point to another. Metals are good conductors of electricity because they allow for the flow of electrons. Rubber is a poor conductor because it does not allow electron movement. Pure water is considered to be a nonconductor. However, when a substance dissolves in water to form ions, the ions are capable of conducting an electric current. If the substance is highly ionized, the solution is a strong conductor. A slightly ionized solute is observed to be a weak conductor. When few, if any, ions are present, the substance is a nonconductor.

An ionic substance dissolves in water to form separate positive and negative ions. Since these opposite charges are electrostatically attracted, it is necessary to reduce the attraction in order for the positive and negative ions to separate. This electrostatic attraction is reduced by the polar water molecules which cluster about each ion. As the electrostatic attraction is reduced, the positive and negative ions separate naturally.

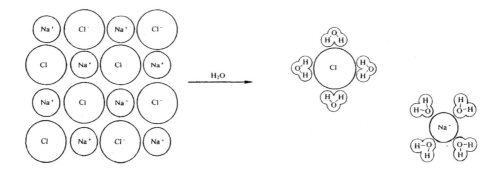

FIGURE 21-1   Dissociation of salt into sodium ions and chloride ions. Notice that the negative end of the polar water molecule is attracted to the positive sodium ion. The positive end of the water molecule is attracted to the negative chloride ion.

Solutions that can conduct an electric current are termed electrolytes and those that cannot are called nonelectrolytes. Electrolytes may be further divided into two groups, strong electrolytes and weak electrolytes. Table 21-1 lists common examples of the three types of electrolytes.

TABLE 21-1 STRONG AND WEAK
ELECTROLYTES AND NONELECTROLYTES

Strong Electrolytes	Weak Electrolytes	Nonelectrolytes
Most soluble salts	Insoluble salts	Sugar
Strong acids	Most acids	Alcohol
(HCl, HNO_3,	Most bases	Water
H_2SO_4, $HClO_4$)		
Strong bases		
(LiOH, NaOH, KOH,		
$Ca(OH)_2$, $Sr(OH)_2$,		
$Ba(OH)_2$)		

In this experiment you will be testing conductivity using an apparatus that has two wires serving as electrodes (Figure 21-2). If the electrodes are immersed in a strongly conducting solution, the circuit is completed and the light bulb in the apparatus glows brightly. A weak electrolyte has few ions in solution and produces only a dull glow in the bulb. A nonelectrolyte does not conduct current and hence the bulb does not glow.

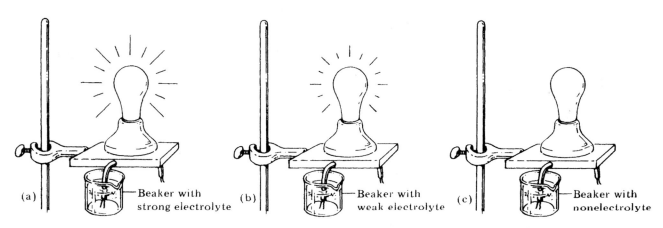

(a) —Beaker with strong electrolyte (b) —Beaker with weak electrolyte (c) —Beaker with nonelectrolyte

FIGURE 21-2   Conductivity apparatus testing (a) a strong electrolyte, (b) a weak electrolyte, and (c) a nonelectrolyte.

Since strong electrolytes are highly ionized, we should indicate the substance in solution as individual ions.

EXAMPLE 21-1   Strong Electrolyte

Aluminum chloride solution is observed to produce a bright glow from the bulb in the conductivity apparatus. Write $AlCl_3$ as it exists in solution.

Solution: Strong conductivity indicates mostly ions in solution, thus, $Al^{3+}_{(aq)} + 3Cl^{1-}_{(aq)}$.

Conversely, weak electrolytes and nonelectrolytes produce few ions in solution and therefore exist primarily in the molecular form.

EXAMPLE 21-2   Weak Electrolyte

Sulfurous acid is observed to produce a dull glow from the bulb in the conductivity apparatus. Portray H_2SO_3 in aqueous solution.

Solution: Weak conductivity indicates mostly undissociated molecules in solution; thus, $H_2SO_{3(aq)}$.

Example 21-3   Nonelectrolyte

A glucose solution is observed to give no glow from the bulb in the conductivity apparatus. Write $C_6H_{12}O_6$ as it exists in solution.

Solution: No conductivity indicates glucose is molecular in solution; thus, $C_6H_{12}O_{6(aq)}$.

Writing Net Ionic Equations

Given the molecular equation for a reaction, balance it by inspection. Change the *molecular equation* into an *ionic equation* using the following guidelines.

1. Each formula of a compound in the molecular equation changes to the *ionic form* if it is a strong electrolyte. Examples include soluble salts, a few acids, and most group IA and IIA hydroxides (bases). The following illustrates changing a formula into ions:

    H_2SO_4 is written $2H^{1+} + SO_4^{2-}$ and $3Ca(NO_3)_2$ is written $3Ca^{2+} + 6NO_3^{1-}$.

2. Each formula of a compound in the molecular equation remains in the *molecular form* if it is a weak electrolyte or nonelectrolyte. Examples include insoluble salts, weak acids and bases, and water.

3. Write the net ionic equation by eliminating those ions not having undergone any change (spectator ions). *Spectator ions are identical* on both sides of the equation. Check the equation for balance.

EXAMPLE 21-4

$$H_2SO_{4(aq)} + 2\,NaOH_{(aq)} \longrightarrow Na_2SO_{4(aq)} + 2\,H_2O_{(\ell)}$$

$$2\,H^{1+}{}_{(aq)} + \cancel{SO_4}^{2-}{}_{(aq)} + 2\,\cancel{Na}^{1+}{}_{(aq)} + 2\,OH^{1-}{}_{(aq)} \longrightarrow 2\,\cancel{Na}^{1+}{}_{(aq)} + \cancel{SO_4}^{2-}{}_{(aq)} + 2\,H_2O_{(\ell)}$$

$$H^{1+}{}_{(aq)} + OH^{1-}{}_{(aq)} \longrightarrow H_2O_{(\ell)}$$

EXAMPLE 21-5

$$CaCl_{2(aq)} + K_2CO_{3(aq)} \longrightarrow CaCO_{3(s)} + 2\,KCl_{(aq)}$$

$$Ca^{2+}{}_{(aq)} + 2\,\cancel{Cl}^{1-}{}_{(aq)} + 2\,K^{1+}{}_{(aq)} + CO_3^{2-}{}_{(aq)} \longrightarrow CaCO_{3(s)} + 2\,\cancel{K}^{1+}{}_{(aq)} + 2\,\cancel{Cl}^{1-}{}_{(aq)}$$

$$Ca^{2+}{}_{(aq)} + CO_3^{2-}{}_{(aq)} \longrightarrow CaCO_{3(s)}$$

EXAMPLE 21-6

$$HCl_{(aq)} + NaC_2H_3O_{2(aq)} \longrightarrow NaCl_{(aq)} + HC_2H_3O_{2(aq)}$$

$$H^{1+}{}_{(aq)} + \cancel{Cl}^{1-}{}_{(aq)} + \cancel{Na}^{1+}{}_{(aq)} + C_2H_3O_2^{1-}{}_{(aq)} \longrightarrow \cancel{Na}^{1+}{}_{(aq)} + \cancel{Cl}^{1-}{}_{(aq)} + HC_2H_3O_{2(aq)}$$

$$H^{1+}{}_{(aq)} + C_2H_3O_2^{1-}{}_{(aq)} \longrightarrow HC_2H_3O_{2(aq)}$$

EXAMPLE 21-7

$$2 \, H_3PO_{4(aq)} + 3 \, Ba(OH)_{2(aq)} \longrightarrow Ba_3(PO_4)_{2(s)} + 6 \, H_2O_{(\ell)}$$

$$2 \, H_3PO_{4(aq)} + 3 \, Ba^{2+}_{(aq)} + 6 \, OH^{1-}_{(aq)} \longrightarrow Ba_3(PO_4)_{2(s)} + 6 \, H_2O_{(\ell)}$$

No spectator ions: net ionic and total ionic equations are the same.

EXAMPLE 21-8

$$2 \, Al(NO_3)_{3(aq)} + 3 \, MgCl_{2(aq)} \longrightarrow 2 \, AlCl_{3(aq)} + 3 \, Mg(NO_3)_{2(aq)}$$

$$2 \, \cancel{Al}^{3+}_{(aq)} + 6 \, \cancel{NO_3}^{1-}_{(aq)} + 3 \, \cancel{Mg}^{2+}_{(aq)} + 6 \, \cancel{Cl}^{1-}_{(aq)} \longrightarrow$$
$$2 \, \cancel{Al}^{3+}_{(aq)} + 6 \, \cancel{Cl}^{1-}_{(aq)} + 3 \, \cancel{Mg}^{2+}_{(aq)} + 6 \, \cancel{NO_3}^{1-}_{(aq)}$$

No reaction: spectator ions only.

EQUIPMENT AND MATERIALS

Equipment
- conductivity apparatus
- small, dry beakers (6)
- straw (or 20 cm length of firepolished glass tubing)

A. Chemicals
- glacial acetic acid, concentrated $HC_2H_3O_2$
- sodium chloride, solid NaCl
- sucrose, solid $C_{12}H_{22}O_{11}$
- aluminum nitrate, 0.1 M $Al(NO_3)_3$
- barium chloride, 0.1 M $BaCl_2$
- copper (II) sulfate, 0.1 M $CuSO_4$
- ethyl alcohol, 0.1 M C_2H_5OH

- hydrochloric acid, 0.1 M HCl
- carbonic acid, 0.1 M H_2CO_3
- magnesium hydroxide, saturated $Mg(OH)_2$
- nitric acid, 0.1 M HNO_3
- potassium chromate, 0.1 M K_2CrO_4
- potassium hydroxide, 0.1 M KOH
- sodium hydroxide, 0.1 M NaOH
- sulfuric acid, 0.1 M H_2SO_4

B. Chemicals
- acetic acid, 0.1 M $HC_2H_3O_2$
- ammonia water, 0.1 M $NH_3 \cdot H_2O$
- potassium nitrate, 0.1 M KNO_3
- sodium carbonate, 0.1 M Na_2CO_3
- barium hydroxide, 0.1 M $Ba(OH)_2$

PROCEDURE

A. Conductivity Testing—Evidence of Ions

1. Test first the conductivity of distilled water and then tap water.

   NOTE: After the conductivity test of each chemical, rinse the electrodes with distilled water. Record your observations in the Data Table. Conclude whether the observations indicate a strong electrolyte, weak electrolyte, or nonelectrolyte. Write the chemical species in either ionic or molecular form.

2. Pour about 10 mL of concentrated glacial acetic acid into a dry beaker and test the conductivity. Add several milliliters of distilled water *slowly* to the acid while continuously testing the conductivity.

3. Place a gram of solid sodium chloride in a *dry* beaker and test for conductivity. Add distilled water to the solid salt and retest the conductivity.

4. Put a gram of solid sucrose in a *dry* beaker and test for conductivity. Add distilled water to the solid sugar and retest the conductivity.

5. Test the conductivity of each of the following solutions:
   (a) aluminum nitrate, 0.1 M $Al(NO_3)_3$
   (b) barium chloride, 0.1 M $BaCl_2$
   (c) copper(II) sulfate, 0.1 M $CuSO_4$
   (d) ethyl alcohol, 0.1 M C_2H_5OH
   (e) hydrochloric acid, 0.1 M HCl
   (f) carbonic acid, 0.1 M H_2CO_3
   (g) magnesium hydroxide, saturated $Mg(OH)_2$
   (h) nitric acid, 0.1 M HNO_3
   (i) potassium chromate, 0.1 M K_2CrO_4
   (j) potassium hydroxide, 0.1 M KOH
   (k) sodium hydroxide, 0.1 M $NaOH$
   (l) sulfuric acid, 0.1 M H_2SO_4

B. Conductivity Testing—Evidence of Reaction

1. Test separately the conductivity of 10 mL of 0.1 M $HC_2H_3O_2$ and 10 mL of 0.1 M $NH_3 \cdot H_2O$. Pour the solutions together and retest the conductivity. Record your observations and conclusions; write balanced molecular, total ionic, and net ionic equations.
2. Add 1 mL of 0.1 M H_2SO_4 into a beaker containing 25 mL of water. Separately test the conductivity of H_2SO_4 and 0.1 M $Ba(OH)_2$. Continuously test the conductivity of the H_2SO_4 solution while adding 2 mL of $Ba(OH)_2$ dropwise. Record your observations and conclusions; write balanced molecular, total ionic, and net ionic equations.
3. Add 1 mL of 0.1 M $Ba(OH)_2$ into a beaker containing 25 mL of water. Test the conductivity. Allow the electrodes to remain in solution and blow through a straw into the solution until the conductivity is a minimum. Exhaling carbon dioxide into water produces carbonic acid, H_2CO_3. Record your observations and conclusions; write balanced molecular, total ionic, and net ionic equations.
4. Test separately the conductivity of 10 mL of 0.1 M Na_2CO_3 and 10 mL of 0.1 M KNO_3. Add the solutions together and retest the conductivity. Record your observations and conclusions; write balanced molecular, total ionic, and net ionic equations.

C. Net Ionic Equations—A Study Assignment

Balance the following molecular equations and write the total and net ionic equations. See the section, Writing Net Ionic Equations, for directions and examples.

1. $H_2SO_{4(aq)} + NH_3 \cdot H_2O_{(aq)} \longrightarrow (NH_4)_2SO_{4(aq)} + H_2O_{(\ell)}$
2. $FeCl_{3(aq)} + Mg(NO_3)_{2(aq)} \longrightarrow MgCl_{2(aq)} + Fe(NO_3)_{3(aq)}$
3. $Ba(C_2H_3O_2)_{2(aq)} + Na_2SO_{4(aq)} \longrightarrow BaSO_{4(s)} + NaC_2H_3O_{2(aq)}$
4. $K_2CO_{3(aq)} + HCl_{(aq)} \longrightarrow KCl_{(aq)} + H_2O_{(\ell)} + CO_{2(g)}$

PRELABORATORY ASSIGNMENT*

1. In your own words define the following terms: molecular equation, net ionic equation, nonelectrolyte, spectator ions, strong electrolyte, total ionic equation, weak electrolyte.
2. Explain the meaning of the symbols: (g), (ℓ), (s), (aq).
3. Give three examples for each of the following:
   (a) strong electrolyte
   (b) weak electrolyte
   (c) nonelectrolyte
4. What will be observed when conductivity-testing a
   (a) strong electrolyte?
   (b) weak electrolyte?
   (c) nonelectrolyte?
5. Write the following as they primarily exist in aqueous solution, that is, ionic form or molecular form.
   (a) $KCl_{(aq)}$—a strong electrolyte
   (b) $C_3H_5(OH)_{3(aq)}$—a nonelectrolyte
   (c) $HNO_{2(aq)}$—a weak electrolyte
6. Why must the electrodes on the conductivity apparatus, as well as all beakers, be rinsed with *distilled* water?
7. What safety precautions should be observed in this experiment?

*Answers in Appendix II.

B. Conductivity Testing—Evidence of Reaction

Substance	Observation	Conclusion
1. $HC_2H_3O_{2(aq)}$		
$NH_3 \cdot H_2O_{(aq)}$		
$HC_2H_3O_2 + NH_3 \cdot H_2O$		
molecular:	$HC_2H_3O_{2(aq)} + NH_3 \cdot H_2O_{(aq)} \longrightarrow NH_4C_2H_3O_{2(aq)} + H_2O_{(\ell)}$	
total ionic:		
net ionic:		
2. $H_2SO_{4(aq)}$		
$Ba(OH)_{2(aq)}$		
$H_2SO_4 + Ba(OH)_2$		
molecular:	$H_2SO_{4(aq)} + Ba(OH)_{2(aq)} \longrightarrow BaSO_{4(s)} + H_2O_{(\ell)}$	
total ionic:		
net ionic:		
3. $Ba(OH)_{2(aq)}$		
$Ba(OH)_2 + H_2CO_3$		
molecular:	$Ba(OH)_{2(aq)} + H_2CO_{3(aq)} \longrightarrow BaCO_{3(s)} + H_2O_{(\ell)}$	
total ionic:		
net ionic:		
4. $Na_2CO_{3(aq)}$		
$KNO_{3(aq)}$		
$Na_2CO_3 + KNO_3$		
molecular:	$Na_2CO_{3(aq)} + KNO_{3(aq)} \longrightarrow K_2CO_{3(aq)} + NaNO_{3(aq)}$	
total ionic:		
net ionic:		

C. Net Ionic Equations—A Study Assignment

1. molecular: $H_2SO_{4(aq)} + NH_3 \cdot H_2O_{(aq)} \longrightarrow (NH_4)_2SO_{4(aq)} + H_2O_{(\ell)}$

   total ionic:

   net ionic:

2. molecular: $FeCl_{3(aq)} + Mg(NO_3)_{2(aq)} \longrightarrow MgCl_{2(aq)} + Fe(NO_3)_{3(aq)}$

   total ionic:

   net ionic:

3. molecular: $Ba(C_2H_3O_2)_{2(aq)} + Na_2SO_{4(aq)} \longrightarrow BaSO_{4(s)} + NaC_2H_3O_{2(aq)}$

   total ionic:

   net ionic:

4. molecular: $K_2CO_{3(aq)} + HCl_{(aq)} \longrightarrow KCl_{(aq)} + H_2O_{(\ell)} + CO_{2(g)}$

   total ionic:

   net ionic:

POSTLABORATORY ASSIGNMENT   NAME _____

1. Explain why distilled water is a nonelectrolyte and tap water is a weak electrolyte. (Procedure A.1)

   _____

2. Why does concentrated glacial acetic acid act first as a nonconductor and then behave as a weak conductor with the addition of water (another nonconductor)? (A.2)

   _____

3. Why does solid sodium chloride act as a nonelectrolyte while the aqueous solution acts as a strong electrolyte? (A.3)

   _____

4. Study the net ionic equation for $HC_2H_3O_2$ and $NH_3 \cdot H_2O$ and explain why the solutions are weak electrolytes individually but a strong conductor when mixed. (B. 1)

   _____

5. Study the net ionic equation for H_2SO_4 and $Ba(OH)_2$ and explain why the solutions conduct individually but not together. (B.2)

   _____

6. Barium hydroxide, $Ba(OH)_2$, acts as a strong electrolyte. Passing carbon dioxide gas through the solution reduces the conductivity to that of a nonelectrolyte. Explain this observation; refer to the net ionic equation. (B.3)

   _____

7. Does the net ionic equation in Procedure B.4 agree with the criteria for chemical reaction, that is, precipitate, gas, etc?

8. State whether each of the following is a: strong electrolyte (75 to 100% ionized), weak electrolyte (1 to 5 % ionized), or nonelectrolyte (0 to 1% ionized).

   (a) $CuS_{(s)}$

   (b) $Sr(OH)_{2\,(aq)}$

   (c) $C_3H_7OH_{(\ell)}$

   (d) $NaHCO_{3\,(aq)}$

9. Balance the molecular equations below and then write the total ionic and net ionic equations. Designate the state of each species by (s), (ℓ), (g), or (aq).

   (a) $ZnCl_{2\,(aq)} + Na_2CO_{3(aq)} \longrightarrow ZnCO_{3(s)} + NaCl_{(aq)}$

   total ionic:

   net ionic:

   (b) $AlCl_{3\,(aq)} + NH_3 \cdot H_2O_{(aq)} \longrightarrow Al(OH)_{3(s)} + NH_4Cl_{(aq)}$

   total ionic:

   net ionic:

   (c) $HC_2H_3O_{2\,(aq)} + Ca(OH)_{2\,(aq)} \longrightarrow Ca(C_2H_3O_2)_{2\,(aq)} + H_2O_{(\ell)}$

   total ionic:

   net ionic:

10. (optional) Define the difference between the terms ionization and dissociation. Give an example of each.

Oxidation-Reduction

OBJECTIVES

1. To observe the oxidation numbers for an element in a compound or ion.
2. To be able to write balanced chemical equations for redox reactions.
3. To determine the electromotive series for several metals.

DISCUSSION

An *oxidation number* is the positive or negative whole number that describes the combining capacity of an element in a compound or ion. By convention, elements are assigned a value of zero when present in an uncombined form. For example, zinc and oxygen would each have a value of zero in the free state. In the compound zinc oxide (ZnO), the oxidation number for zinc is a positive two (Zn^{2+}) and for oxygen a negative two (O^{2-}).

PROBLEM EXAMPLE 22-1

Calculate the oxidation number (ox no) of carbon in the compound sodium hydrogen carbonate, $NaHCO_3$.

Solution: The oxidation number of sodium is +1, hydrogen is +1, and oxygen is —2. Since the sum of all oxidation numbers for the elements in a compound must equal zero, we can write the equation

$$+1 +1 + \text{ox no C} + 3\,(-2) = 0$$
$$+2 + \text{ox no C} - 6 = 0$$
$$\text{ox no C} = +4$$

In calculating an oxidation number we will use the following rules:

- The oxidation number of an element in the free state is zero.
- For a compound, the sum of the oxidation numbers for all atoms is zero.
- For an ion, the sum of the oxidation numbers for all atoms is equal to the ionic charge.
- In general, the oxidation number of hydrogen is +1 and oxygen is —2.*
- In general, the oxidation number of metals in Group IA is +1; Group IIA is +2, and Group IIIA is +3.*

*There are a few exceptions to these general rules. However, none of the exceptions appear within this experiment.

PROBLEM EXAMPLE 22-2

Calculate the oxidation number of carbon in the oxalate ion, $C_2O_4{}^{2-}$.

Solution: The oxidation number of each oxygen atom is -2. The polyatomic ion has a charge of $2-$ and there are two carbon atoms; therefore,

$$2(\text{ox no C}) + 4\ (-2) = -2$$
$$2(\text{ox no C}) - 8 = -2$$
$$2(\text{ox no C}) = -2 + 8 = +6$$
$$\text{ox no C} = +3$$

Oxidation-reduction equations can be balanced by two methods: the *oxidation number method* or the *ion electron method*. The following example illustrates the oxidation number method.

PROBLEM EXAMPLE 22-3

Balance the following equation by the oxidation number method. The reaction takes place in acidic solution.

$$Ag + NO_3{}^{1-} \longrightarrow Ag^{1+} + NO$$

Solution: First, note the change in oxidation number for silver: 0 to +1. Silver loses $1e^-$. Second, note the change in oxidation number for nitrogen: +5 to +2. Nitrogen gains $3e^-$.

To balance electrons, the coefficient of silver is three; thus, 3 Ag and $3\ Ag^{1+}$. The coefficient of each nitrogen species is one; thus, $NO_3{}^{1-}$ and NO.

$$3\ Ag + NO_3{}^{1-} \longrightarrow 3\ Ag^{1+} + NO$$

The next step is to balance the oxygen atoms using water molecules:

$$3\ Ag + NO_3{}^{1-} \longrightarrow 3\ Ag^{1+} + NO + 2\ H_2O$$

Since the reaction takes place in acidic solution we will balance the hydrogen atoms using hydrogen ions:

$$4\ H^{1+} + 3\ Ag + NO_3{}^{1-} \longrightarrow 3\ Ag^{1+} + NO + 2\ H_2O$$

The last step is to check the equation for balance. A redox equation must be ~~balanced so that the number of atoms~~ ... ... of the reactants and products are equal.

$$4\ H^{1+} + 3\ Ag + NO_3{}^{1-} \longrightarrow 3\ Ag^{1+} + NO + 2H_2O$$

The check reveals that the atoms of reactants and products are balanced. Furthermore, the total ionic charge of the reactants is +3; the total product charge is also +3.

The *ion electron method* of balancing redox equations is also termed the *half-reaction method*. The oxidation reaction is treated separately from the reduction reaction although each can only take place in conjunction with the other. The two half-reactions are then added together to give the balanced redox equation.

PROBLEM EXAMPLE 22-4

Balance the reaction in the previous example using the half-reaction method.

$$Ag + NO_3^{1-} \longrightarrow Ag^{1+} + NO \text{ (in acid)}$$

Solution: First, write the partial equation for each half-reaction:

$$Ag \longrightarrow Ag^{1+}$$

$$NO_3^{1-} \longrightarrow NO$$

Balance the partial equation using water and hydrogen ions:

$$Ag \longrightarrow Ag^{1+}$$

$$4 H^{1+} + NO_3^{1-} \longrightarrow NO + 2 H_2O$$

Next, write the two half-reactions and balance the charge using electrons:

Oxidation: $\qquad\qquad Ag \longrightarrow Ag^{1+} + 1e^-$

Reduction: $\qquad 4 H^{1+} + NO_3^{1-} + 3e^- \longrightarrow NO + 2 H_2O$

The reduction half-reaction gains $3e^-$ for every $1e^-$ lost in the oxidation half-reaction. Therefore, we will multiply the oxidation half-reaction by three:

$$3 Ag \longrightarrow 3 Ag^{1+} + 3e^-$$

Let's add the two half-reactions together to obtain the redox equation:

$$3 Ag + 4 H^{1+} + NO_3^{1-} + 3e^- \longrightarrow NO + 2 H_2O + 3 Ag^{1+} + 3e^-$$

We can simplify the equation by cancelling $3e^-$ on each side. Finally, let's check the equation for balance.

$$3 Ag + 4 H^{1+} + NO_3^{1-} \longrightarrow NO + 2 H_2O + 3 Ag^{1+}$$

The number of atoms of reactants and products are equal. Also notice that the total ionic charge is +3 for both reactants and products.

The *electromotive series* is a list of metals arranged in order of ability to displace another metal from its aqueous solution. Metals listed first in the series are more reactive and thus will reduce a less active cation to the metal in the free state. For example, mercury is more active than gold; hence

$$Hg_{(\ell)} + Au^{3+}_{(aq)} \longrightarrow Hg^{2+}_{(aq)} + Au_{(s)}$$

On the other hand, the opposite reaction is denied because gold is lower than mercury in the electromotive series.

$$Au_{(s)} + Hg^{2+}_{(aq)} \longrightarrow \text{No Reaction}$$

For reference purposes, hydrogen (H) is included in the series. Metals above hydrogen in the electromotive series displace hydrogen gas from acid solutions. Metals below hydrogen do not react with dilute acids.

PROBLEM EXAMPLE 22-5

An iron nail develops a shiny metallic surface when placed into a cadmium solution. The equation is:

$$Fe_{(s)} + Cd^{2+} \longrightarrow Fe^{2+} + Cd_{(s)}$$

The shiny metallic cadmium reacts with dilute acid according to the equation:

$$Cd_{(s)} + 2\ H^{1+} \longrightarrow Cd^{2+} + H_{2(g)}$$

However, liquid mercury does not react with dilute acid.

$$Hg_{(\ell)} + H^{1+} \longrightarrow No\ Reaction$$

Deduce the electromotive series for cadmium, iron, mercury, and hydrogen.

Solution: Since iron displaces cadmium from aqueous solution, it must be higher in the series (Fe > Cd).

Cadmium displaces hydrogen gas from acid solution; thus the metal is above hydrogen in the series (Cd > H). However, mercury does not react with dilute acid and is therefore below hydrogen in the series (H > Hg).

In summary, the observed electromotive series is:

Fe	Cd	(H)	Hg
most active			*least active*

EQUIPMENT AND CHEMICALS

Equipment

- test tubes (6) with rack
- dropper pipets for solutions
- evaporating dish

Chemicals

- manganese, Mn metal
- potassium permanganate, 0.010 M $KMnO_4$
- sodium sulfite, 0.5 M Na_2SO_3
- dilute hydrochloric acid, 6 M HCl
- chromium, Cr metal
- potassium dichromate, 0.1 M $K_2Cr_2O_7$
- dilute sodium hydroxide, 6 M NaOH
- sulfur, S powder

- sodium sulfide, solid Na_2S
- dilute sulfuric acid, 6 M H_2SO_4
- iodine solution, 0.5 M I_2/KI
- sodium thiosulfate, 0.1 M $Na_2S_2O_3$
- dilute nitric acid, 6 M HNO_3
- copper, Cu metal
- concentrated nitric acid, 16 M HNO_3
- ammonium chloride, solid NH_4Cl
- dilute sodium hydroxide, 6 M NaOH
- lead, Pb metal
- magnesium, Mg metal
- zinc, Zn metal
- zinc sulfate, 0.1 M $ZnSO_4$
- silver nitrate, 0.1 M $AgNO_3$
- unknown metal samples

PROCEDURE

A. Oxidation Numbers of Manganese

1. Examine a piece of manganese metal. Record your observations in the Data Table and state the oxidation number.
2. Deliver 2 mL (1/10 test tube) of potassium permanganate, $KMnO_4$, into a test tube. Note the color and calculate the oxidation number of manganese. Add a few drops of dilute sodium hydroxide and one drop of sodium sulfite into the test tube. Observe the color change for the formation of manganate ion, MnO_4^{2-}. Calculate the oxidation number of manganese in this ion.

3. Introduce 2 mL of potassium permanganate solution into a clean test tube. Add sodium sulfite drop by drop until the purple color fades. After several minutes, observe the solid particles of manganese dioxide, MnO_2, forming in the solution. Calculate the oxidation number of manganese.

4. Place 2 mL of potassium permanganate into a clean test tube. Add a drop of dilute hydrochloric acid and then several drops of sodium sulfite. Record your observation and state the oxidation number for the new manganese species, Mn^{2+}.

B. Oxidation Numbers of Chromium

1. Inspect chromium metal and record the oxidation number in the Data Table.

2. Deliver 2 mL of aqueous potassium dichromate, $K_2Cr_2O_7$, into a test tube. Note the color and calculate the oxidation number of chromium. Add a few drops of dilute sodium hydroxide into the test tube and observe the color change. Calculate the oxidation number of chromium in the chromate ion, CrO_4^{2-}, that was produced.

3. Put 2 mL of potassium dichromate solution into a test tube. Add one drop of dilute hydrochloric acid and 2 mL of sodium sulfite solution; record the color change. Calculate the oxidation number for the chromium ion, Cr^{3+}, that was produced.

C. Oxidation Numbers of Sulfur

1. Observe powdered sulfur and record its oxidation number.

2. Place a pea-sized portion of sulfur in an evaporating dish and ignite it with a burner. Notice the color of the flame and the white sulfur dioxide gas. Find the oxidation number of sulfur in gaseous SO_2.

   SAFETY: Perform this operation under a fume hood. Avoid breathing the sulfur dioxide gas.

3. Put a very small crystal of sodium sulfide, Na_2S, into a dry test tube and several drops of dilute sulfuric acid, H_2SO_4. Describe the odor of the hydrogen sulfide, H_2S, gas given off. Calculate the oxidation number of sulfur in Na_2S, H_2SO_4, and H_2S.

   SAFETY: Note the odor of the hydrogen sulfide carefully. Avoid breathing an excess and dispose of the test tube contents under a fume hood.

4. Pour about 2 mL of iodine solution into a test tube. Add sodium thiosulfate, $Na_2S_2O_3$, until the iodine is discolored. Record the change and calculate the oxidation number for sulfur in $Na_2S_2O_3$.

D. Oxidation Numbers of Nitrogen

1. Air contains over 78 percent nitrogen. By simply breathing the air, note the color, odor, and taste. State the oxidation number for nitrogen, N_2, in the air.

2. Put 2 mL of dilute nitric acid, HNO_3, into a test tube. Add one small piece of copper and observe the reaction. Calculate the oxidation number of nitrogen in the gas evolved; the gas is nitrogen monoxide, NO.

3. Deliver 2 mL of *concentrated* nitric acid, HNO_3, into a clean test tube. Observe the color of the acid solution and calculate the oxidation number for nitrogen. Add a small piece of copper metal and observe the color of the nitrogen dioxide, NO_2, gas produced. Find the oxidation number for nitrogen.

   SAFETY: *Concentrated* nitric acid should be handled with great caution. Avoid breathing the nitrogen dioxide gas and dispose of the test tube contents under a fume hood.

4. Using a scoopula, introduce a pea-sized portion of solid ammonium chloride, NH_4Cl, into a clean test tube. Add a dropper of dilute sodium hydroxide and cover the end of the test tube with your thumb for 30 seconds. Shake the test tube, release your thumb, and carefully observe any odor. Find the oxidation number of nitrogen in the ammonia gas, NH_3, released. Also, calculate the oxidation number of nitrogen in NH_4Cl.

E. **Oxidation-Reduction Equations**   Many of the reactions in Procedures A-D illustrate oxidation-reduction and are listed in the Data Table. Balance each redox equation using either the oxidation number method or ion electron (half-reaction) method.

F. **Electromotive Series of an Unknown Metal**

1. Obtain an unknown metal *(M)* and record the number in the Data Table.
2. Add 2 mL of dilute hydrochloric acid into each of four test tubes. Put a small piece of Cu, Mg, Zn, or *M* into each test tube. Record your observations in the Data Table.
3. Clean the four test tubes and then add 2 mL of zinc sulfate solution. Put a small piece of Cu, Mg, Zn, or *M* into each test tube. Observe and record the reactions.
4. Clean the four test tubes and then add 2 mL of silver nitrate solution. Put a small piece of Cu, Mg, Zn, or *M* into each test tube. Record your observations.
5. Based upon the foregoing observations, list an experimental electromotive series for the following: Cu, Mg, Zn, Ag, (H), and *M*.

PRELABORATORY ASSIGNMENT*

1. In your own words define the following terms: electromotive series, oxidation, oxidation number, oxidizing agent, redox reaction, reducing agent, reduction.
2. Calculate the oxidation number of chlorine in (a) Cl_2; (b) HCl; (c) NaClO; (d) ClO_4^{1-}; and (e) Cl_2O_5.
3. Balance the following oxidation-reduction equations:
   (a) $Cr_2O_7^{2-} + Fe^{2+} + H^{1+} \longrightarrow Cr^{3+} + Fe^{3+} + H_2O$
   (b) $Mn^{2+} + H_2O_2 + OH^{1-} \longrightarrow MnO_2 + H_2O$
4. Assume the following reactions go essentially to completion:
$$Al + Cd^{2+} \longrightarrow Al^{3+} + Cd$$
$$Ni + Ag^{1+} \longrightarrow Ni^{2+} + Ag$$
$$Cd + Ni^{2+} \longrightarrow Cd^{2+} + Ni$$
   Establish the electromotive series for the metals Al, Cd, Ni, and Ag. List the most active metal first.
5. What safety precautions must be observed in this experiment?

*Answers in Appendix II.

1. Calculate the oxidation number for iodine in each of the following.

   (a) KIO_4 _____+7_____
       +1 -8

   (b) HI   _____-1_____
       +1 -1

   (c) IF_7 _____+7_____
       +7 -7

   (d) NaIO   _____+1_____
       +1 -2

   (e) IO_2^{1-} _____+3_____
       +3 -4

   (f) I_2O_5 _____+5_____
       +5 -10

   (g) IO_3^{1-} _____+5_____
       +5 -6

   (h) I_2 _____0_____

2. Balance the following oxidation-reduction equations. Assume each reaction takes place in acidic solution.

   (a) $MnO_4^{1-} + Fe^{2+} \longrightarrow Mn^{2+} + Fe^{3+}$
         +7 -8

   (b) $Sn^{2+} + IO_3^{1-} \longrightarrow Sn^{4+} + I^{1-}$

   (c) $NO_3^{1-} + H_2S \longrightarrow NO + S$

   (d) $Mn^{2+} + BiO_3^{1-} \longrightarrow MnO_4^{1-} + Bi^{3+}$

   (e) $Cr_2O_7^{2-} + C_2O_4^{2-} \longrightarrow Cr^{3+} + CO_2$

3. Balance the following oxidation-reduction equations. Assume each reaction takes place in basic solution.

(a) $SO_3^{2-} + Cl_2 \longrightarrow SO_4^{2-} + Cl^{1-}$

(b) $MnO_2 + O_2 \longrightarrow MnO_4^{2-} + H_2O$

(c) $MnO_4^{1-} + BrO_2^{1-} \longrightarrow MnO_2 + BrO_4^{1-}$

4. Consider the following redox equation:

$I_2 + 5 Cl_2 + 6 H_2O \longrightarrow 2 IO_3^{1-} + 10 Cl^{1-} + 12 H^{1+}$

+5,5-12

Identify the species undergoing

(a) oxidation ___I___   (b) reduction ___Cl___

Identify the species functioning as

(a) the oxidizing agent ___I___   (b) the reducing agent ___Cl O___

5. Solutions of bromide, fluoride, and iodide were treated with bromine water. Iodide was oxidized to iodine; bromide and fluoride did not react. Arrange the three elements Br_2, F_2, and I_2 in order of their ability to oxidize another species.

Strongest oxidizing agent: _____   _____   _____

6. (optional) Certain species demonstrate the ability to undergo oxidation and reduction simultaneously. This phenomenon is termed disproportionation. For example, chlorine in basic solution produces chlorate and chloride ions. Write a balanced redox equation for the reaction.

Oxidation # decreases = oxidizing agent
oxidation # increases = reducing agent

Organic Model Exercise

OBJECTIVES

1. To build example model structures of the aliphatic hydrocarbons: alkanes, alkenes, alkynes.
2. To build example model structures of the aromatic hydrocarbons.
3. To build example model structures of the hydrocarbon derivatives: organic halides, alcohols, phenols, ethers, aldehydes, ketones, carboxylic acids, esters, amides, amines.
4. To identify the class of compound represented by the functional group in unknown model structures.
5. To acquire a three-dimensional perspective of organic compounds from the building of molecular models.

DISCUSSION

Organic chemistry is the study of compounds that contain the element carbon; compounds that do not contain carbon are termed inorganic. It is interesting to note that the element carbon is found in over two million different compounds. On the other hand, about 100,000 identified compounds do not contain carbon in their chemical formula. There are two reasons why over 90 percent of all compounds are organic. First, carbon is unusual in that it has the ability to self-link, forming chains of carbon atoms of varying length. Second, organic compounds typically contain several carbon atoms that may be joined together in more than one arrangement or configuration.

Compounds having the same molecular formula but a different configuration are termed *isomers*. For instance, the molecular formula C_4H_{10} may be constructed in two ways and still satisfy the bond requirements for carbon (four bonds) and hydrogen (one bond). Figure 23-1 illustrates the isomers of butane, C_4H_{10}.

Although the entire molecule as well as the individual bonds can be rotated in space to give what appear to be additional configurations for the formula C_4H_{10}, careful examination will reveal there are only two possibilities.

Although there are over two million organic compounds, they may be dealt with effectively by systematic classification. *Hydrocarbons* may be classified as alkanes, alkenes, alkynes, or aromatic. *Aromatic hydrocarbons* contain a benzene nucleus within the compound. If the hydrocarbon contains a *functional group*, such as an alcohol or ketone, it is considered a *hydrocarbon derivative*. Figure 23-2 illustrates an overall classification scheme.

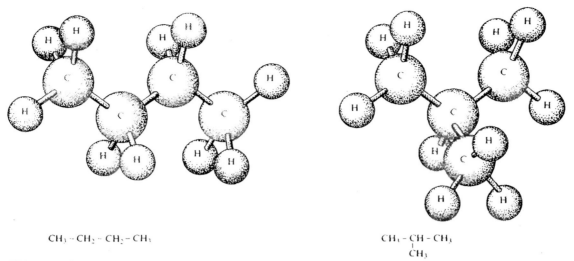

CH₃ - CH₂ - CH₂ - CH₃   CH₃ - CH - CH₃
                                                  |
                                                 CH₃

FIGURE 23-1 The two isomers of butane which share the molecular formula C_4H_{10}. The ball-and-stick model is shown above; the condensed structural formula below.

In this experiment we will build example molecular models for each of the hydrocarbons and derivatives shown in Figure 23-2. The following examples will serve to correlate the classes of compounds with the ball-and-stick models you will construct.

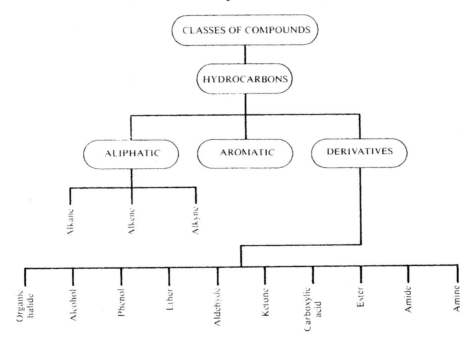

FIGURE 23-2 Classes of organic compounds organized into a systematic classification scheme.

Class of Compound	Example	Model Representation
alkane	$H_3C - CH_3$	
alkene	$H_2C = CH_2$	
alkyne	$HC \equiv CH$	
aromatic	C_6H_6	

Substituting a chlorine, bromine, or iodine atom for a hydrogen atom onto the hydrocarbon chain produces a class of compounds called the organic halides.

organic halide $CH_3 - Cl$

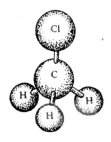

Many organic compounds, especially those of biological interest, contain oxygen as well as carbon and hydrogen. Oxygen has a bond requirement of two, and this may be satisfied in a number of ways. If one bond is attached to a carbon and the second bond to a hydrogen, an alcohol is formed. If both bonds are attached to carbon atoms, an ether is formed.

alcohol $\qquad$ $CH_3CH_2 - OH$

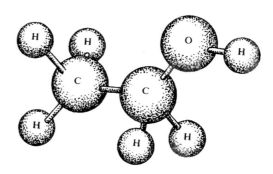

ether $\qquad$ $CH_3 - O - CH_3$

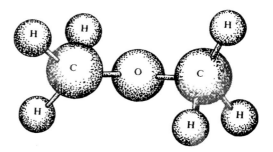

A phenol is simply a special example of an alcohol; that is, the —OH is attached directly to an aromatic group.

Oxygen can also satisfy its bond requirement of two by forming a double bond with carbon; this is termed a *carbonyl group*.

$$\begin{array}{c} O \\ \| \\ - C - \end{array}$$

If the carbonyl group is at the end of the molecule—that is, one of the bonds is attached to a hydrogen—the functional group is an aldehyde. If the carbonyl is in the middle of the carbon chain—that is, both bonds are joined to other carbon atoms—the functional group is a ketone.

aldehyde $\qquad$ $CH_3CH_2 - \overset{\overset{\displaystyle O}{\|}}{C} - H$

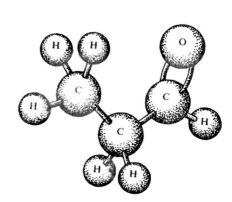

ketone

$$O$$
$$\parallel$$
$$CH_3 - C - CH_3$$

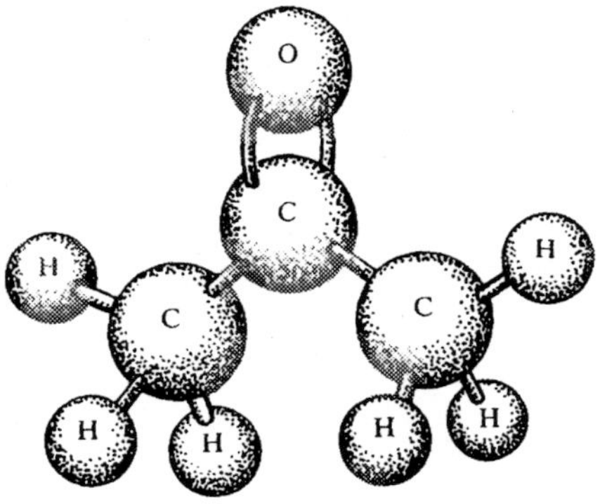

Another possibility is that the carbonyl may be attached to an —OH group. The resulting structure is found in the class of compounds called carboxylic acids.

carboxylic acid

$$O$$
$$\parallel$$
$$CH_3 - C - OH$$

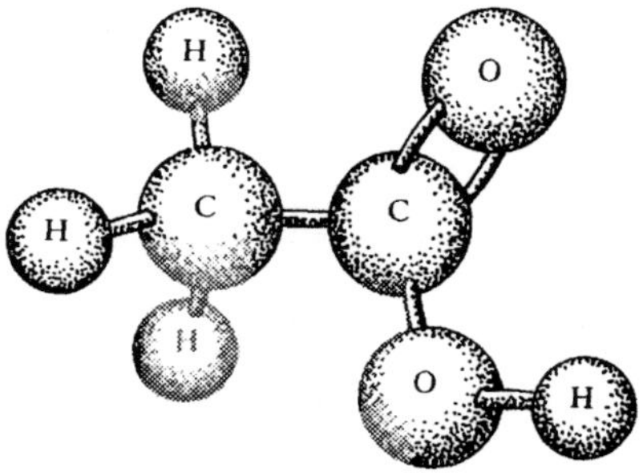

If we substitute a carbon for the hydrogen in the carboxylic acid group, this gives rise to the class of compounds called esters. Esters are noted for their typically fragrant odors.

ester

$$O$$
$$\parallel$$
$$H - C - OCH_3$$

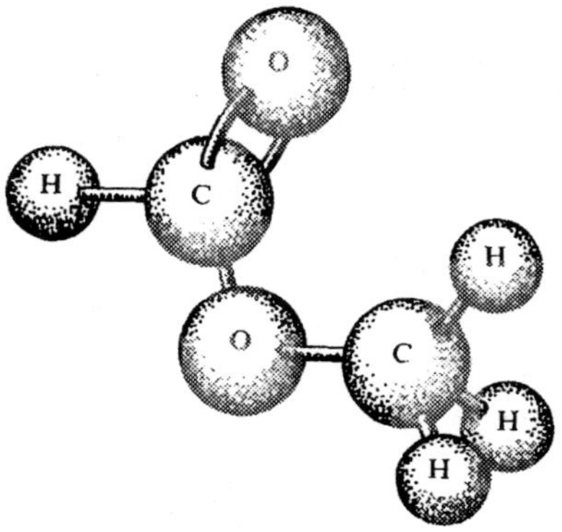

Starting with a carboxylic acid, let's remove the —OH and replace it with —NH₂. The resulting class of compound is called an amide.

amide

$$O$$
$$\parallel$$
$$H - C - NH_2$$

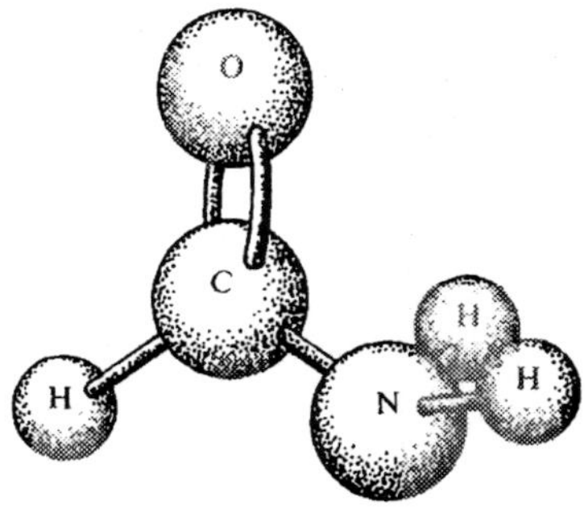

Finally, if we attach the $-NH_2$ group directly to a hydrocarbon, an amine results. Notice that the amine does not contain a carbonyl group.

amine $CH_3 - NH_2$

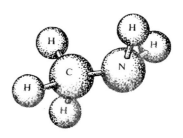

EQUIPMENT AND CHEMICALS

- molecular model kits

Directions for Using Molecular Models. When constructing a model, a hole in a ball represents a missing electron that is necessary in order to complete an octet. If two balls are joined together with a single connector, the connector represents a bond composed of two electrons. If two balls are joined together by two connectors, a double bond is indicated and represents four bonding electrons. Three connectors joining two balls represents a triple bond and a total of six electrons. The six electrons are perhaps more precisely referred to as three pairs of bonding electrons.

one connector — single bond (electron pair)
two connectors — double bond (two electron pairs)
three connectors — triple bond (three electron pairs)

As the model is constructed for a compound, all the holes in each ball should be filled with a connector. (Nitrogen may be an exception.) The color code for each ball is as follows:

white or yellow ball — hydrogen (one hole)
black ball — carbon (four holes)
red ball — oxygen (two holes)
blue ball — nitrogen (four or five holes)
green ball -- chlorine (one hole)
orange ball — bromine (one hole)
purple ball — iodine (one hole)

PROCEDURE

A. Aliphatic Hydrocarbons

1. Construct the molecular models for the following *alkanes* and write their structural formulas in the Data Table:
   (a) methane, CH_4
   (b) ethane, C_2H_6
   (c) propane, C_3H_8
   (d) butane, C_4H_{10}
   (e) isobutane, C_4H_{10}

2. Construct models for the following *alkenes*:
   (a) ethylene, C_2H_4
   (b) propylene, C_3H_6
3. Construct the molecular model for the following *alkynes*:
   (a) acetylene, C_2H_2
   (b) methyl acetylene, C_3H_4

B. Aromatic Hydrocarbons   Construct the molecular models for the following *aromatic* compounds:

   (a) toluene, $C_6H_5CH_3$
   (b) para-xylene, $C_6H_4(CH_3)_2$

C. Derivatives of Hydrocarbons   Construct the molecular models for the following hydrocarbon derivatives:

   1. *Organic Halides*
      (a) chloroform, $CHCl_3$
      (b) carbon tetrachloride, CCl_4
   2. *Alcohols*
      (a) methyl alcohol, CH_3OH
      (b) ethyl alcohol, C_2H_5OH
      (c) propyl alcohol, C_3H_7OH
      (d) isopropyl alcohol, C_3H_7OH
   3. *Phenols*
      (a) phenol, C_6H_5OH
      (b) ortho-cresol, $C_6H_4(CH_3)OH$
   4. *Ethers*
      (a) dimethyl ether, CH_3OCH_3
      (b) diethyl ether, $C_2H_5OC_2H_5$
   5. *Aldehydes*
      (a) formaldehyde, HCHO
      (b) acetaldehyde, CH_3CHO
   6. *Ketones*
      (a) acetone, CH_3COCH_3
      (b) methyl ethyl ketone, $CH_3COC_2H_5$
   7. *Carboxylic Acids*
      (a) formic acid, HCOOH
      (b) acetic acid, CH_3COOH
   8. *Esters*
      (a) methyl formate, $HCOOCH_3$
      (b) ethyl acetate, $CH_3COOC_2H_5$
   9. *Amides*
      (a) formamide, $HCONH_2$
      (b) acetamide, CH_3CONH_2
   10. *Amines*
      (a) methyl amine, CH_3NH_2
      (b) ethyl amine, $C_2H_5NH_2$
      (c) propyl amine, $C_3H_7NH_2$
      (d) isopropyl amine, $C_3H_7NH_2$

D. Identification of Classes of Compounds   The instructor will provide numbered models of unknown organic compounds. Inspect each model, draw the structural formula in the Data Table, circle the functional group, and identify the class of compound to which it belongs. Each compound contains a single functional group. The following classes of compounds may be represented: *alkene, alkyne, organic halide, alcohol, phenol, ether, aldehyde, ketone, carboxylic acid, ester, amide, amine.*

PRELABORATORY ASSIGNMENT*

1. In your own words define the following terms: aromatic hydrocarbons, carbonyl group, functional group, hydrocarbon derivatives, saturated hydrocarbons, structural isomers, unsaturated hydrocarbons.
2. The following colored balls represent an atom of which element: black, red, white, green, orange, purple, blue?
3. What is used to construct a model representing two atoms of carbon joined by a single bond?
4. What is used to construct a model for a carbon atom and oxygen atom joined by a double bond?
5. Indicate the name of the class of compound for each of the following:

(a) $-\overset{\displaystyle |}{\underset{\displaystyle |}{C}}-O-\overset{\displaystyle |}{\underset{\displaystyle |}{C}}-$

(b) $-\overset{\displaystyle |}{\underset{\displaystyle |}{C}}-Cl$

(c) $\overset{\diagdown}{\diagup}C = C\overset{\diagup}{\diagdown}$

(d) $-\overset{\displaystyle |}{\underset{\displaystyle |}{C}}-\overset{\displaystyle O}{\overset{\displaystyle \|}{C}}-\overset{\displaystyle |}{\underset{\displaystyle |}{C}}-$

(e) $-\overset{\displaystyle O}{\overset{\displaystyle \|}{C}}-OH$

(f) $-C \equiv C -$

(g) $-\overset{\displaystyle O}{\overset{\displaystyle \|}{C}}-NH_2$

(h) $-\overset{\displaystyle |}{\underset{\displaystyle |}{C}}-OH$

(i)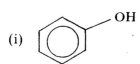

(j) $-\overset{\displaystyle O}{\overset{\displaystyle \|}{C}}-O-\overset{\displaystyle |}{\underset{\displaystyle |}{C}}-$

(k) $-NH_2$

(l) $-\overset{\displaystyle O}{\overset{\displaystyle \|}{C}}-H$

6. State the trivial name for each of the following alkyl groups:

(a) $CH_3 - \overset{\displaystyle |}{CH} - CH_3$

(b) $CH_3 - CH_2 - CH_2 -$

(c) $CH_3 -$

(d) $CH_3 - CH_2 -$

7. Draw and label the ortho(o), meta(m), and para(p) isomers of dichlorobenzene.

DATA TABLE FOR ORGANIC MODEL EXERCISE

A. Aliphatic Hydrocarbons

1. *Alkanes*
   (a) methane, CH_4 (b) ethane, C_2H_6 (c) propane, C_3H_8

   (d) butane, C_4H_{10} (e) isobutane, C_4H_{10}

2. *Alkenes*
   (a) ethylene, C_2H_4 (b) propylene, C_3H_6

3. *Alkynes*
   (a) acetylene, C_2H_2 (b) methyl acetylene, C_3H_4

B. Aromatic Hydrocarbons

   (a)   toluene, $C_6H_5CH_3$ (b) para-xylene, $C_6H_4(CH_3)_2$

C. Derivatives of Hydrocarbons

1. *Organic Halides*
   (a) chloroform, $CHCl_3$ (b) carbon tetrachloride, CCl_4

2. *Alcohols*
   (a) methyl alcohol, CH_3OH (b) ethyl alcohol, C_2H_5OH

   (c) propyl alcohol, C_3H_7OH (d) isopropyl alcohol, C_3H_7OH

3. *Phenols*
   (a) phenol, C_6H_5OH (b) ortho-cresol, $C_6H_4(CH_3)OH$

4. *Ethers*
   (a) dimethyl ether, CH_3OCH_3 (b) diethyl ether, $C_2H_5OC_2H_5$

5. *Aldehydes*
   (a) formaldehyde, HCHO
   (b) acetaldehyde, CH_3CHO

6. *Ketones*
   (a) acetone, CH_3COCH_3
   (b) methyl ethyl ketone, $CH_3COC_2H_5$

7. *Carboxylic Acids*
   (a) formic acid, HCOOH
   (b) acetic acid, CH_3COOH

8. *Esters*
   (a) methyl formate, $HCOOCH_3$
   (b) ethyl acetate, $CH_3COOC_2H_5$

9. *Amides*
   (a) formamide, $HCONH_2$
   (b) acetamide, CH_3CONH_2

10. *Amines*
    (a) methyl amine, CH_3NH_2
    (b) ethyl amine, $C_2H_5NH_2$

    (c) propyl amine, $C_3H_7NH_2$
    (d) isopropyl amine, $C_3H_7NH_2$

D. Identification of Classes of Compounds

Model Number	Structure	Class of Compound
#1		
#2		
#3		
#4		
#5		
#6		
#7		
#8		
#9		
#10		

1. Identify the class of compound (for example, alcohol) corresponding to each of the functional groups circled below.

(a) citric acid (the tart taste in oranges, lemons, and limes)

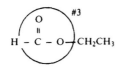

#1 _____

#2 _____

(b) ethyl formate (rum flavor)

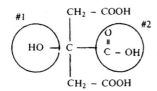

#3 _____

(c) urea (in animal urine)

#4 _____

(d) acetophenone (in tear gas)

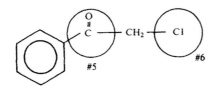

#5 _____

#6 _____

(e) thyroxine (thyroid hormone)

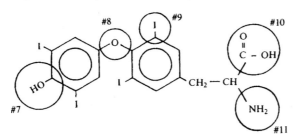

#7 _____

#8 _____

#9 _____

#10 _____

#11 _____

(f) cinnamaldehyde (in cinnamon)

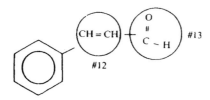

#12 _____

#13 _____

(g) norethinodrone (oral contraceptive
        hormone)

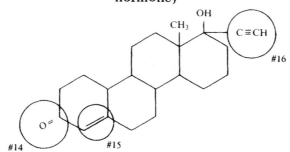

#14 _____

#15 _____

#16 _____

2. Write the structural formula for the four isomers having the molecular formula $C_3H_6Br_2$.

(a)   (b)

(c)   (d)

3. (optional) Give the preferable systematic IUPAC name for each of the following:

(a) ethylene   (b) acetylene

(c) chloroform   (d) propyl alcohol

(e) formaldehyde   (f) acetone

(g) acetic acid   (h) methyl acetate

(i) formamide   (j) isopropyl amine

Appendices

I

Glossary

abscissa   The horizontal axis (x-axis) on a graph.

absolute error   The difference between the experimental value and the theoretical value. A measure of accuracy.

absolute temperature   Temperature expressed in kelvin units.

absolute zero   Temperature at which an ideal gas has no kinetic energy; 0 K or $-273.15°C$.

accuracy   The closeness of an experimental measurement to the theoretical or accepted value.

acid (Arrhenius definition)   A substance that yields hydrogen ions (H^{1+}) when dissolved in water.

acid (Brønsted-Lowry definition)   A substance that can give or donate a proton (H^{1+}) to some other substance.

actinide series   The elements 90 to 103.

active metal   A metal with sufficient potential to react directly with water; includes Li, Na, K, Ca, Sr, and Ba.

actual yield   The amount of product actually obtained from a reaction.

alkali metals   The Group IA elements, excluding hydrogen.

alkaline earth metals   The Group IIA elements.

Angstrom (Å)   A unit of length used to denote atomic dimensions. $1Å = 10^{-8}$ cm.

anhydrous salt   A salt not containing water. A hydrate after losing its water of crystallization.

anions   Ions with a negative charge.

aqueous solution   A solution having water as the solvent.

aromatic hydrocarbons   Hydrocarbons that include benzene and compounds containing aliphatic groups attached to aromatic rings.

atmospheric pressure   The pressure exerted by the gas molecules in the atmosphere; pressure is measured with a barometer.

atomic mass scale   Relative scale of atomic masses, based on arbitrarily assigned value of exactly 12 atomic mass units (amu) for the mass of carbon-12.

atomic mass unit (amu)   The mass corresponding to exactly 1/12 the mass of a carbon-12 atom.

atomic number   The number of protons in the nucleus of an atom for a given element.

Avogadro's number (N) 6.02 × 10²³ The number of particles such as atoms, formula units, molecules, or ions that constitutes one mole of the said particle.

balance An instrument for measuring mass.

barometer A device for measuring the atmospheric pressure.

base (Arrhenius definition) A substance that yields hydroxide ions (OH^{1-}) when dissolved in water.

base (Brønsted-Lowry definition) A substance capable of receiving or accepting a proton (H^{1+}) from some other substance.

calorie The amount of heat required to raise the temperature of 1 g of water from $14.5°C$ to $15.5°C$.

calorimeter A device used to measure changes in heat within a closed system.

carbonyl group The functional group that consists of a carbon double bonded to an oxygen.

catalyst A substance that speeds up a chemical reaction but is recovered without appreciable change at the end of the reaction.

cations Ions with a positive charge.

centrifuging The act of separating a precipitate from solution by spinning in a centrifuge.

change of state A change of physical state; for example, liquid to solid.

chemical changes Changes in substances that can be observed only when a change in the chemical composition of the substance is occurring. New substances are formed.

chemical properties Properties of substances that can be observed only when a substance undergoes a change in chemical composition.

chemistry A study of the composition of matter and the changes it undergoes.

combination reaction Reaction in which two or more elements or compounds combine.

combined gas laws Gas laws combining Charles' Law and Boyle's Law into a single expression. Relates pressure, volume, and temperature.

compound A pure substance that can be broken down by various chemical means into two or more different substances.

concentrated solution A solution containing a large amount of dissolved solute.

conditioning The rinsing of a piece of glassware with the solution to be measured in order to prevent dilution by water on the inside surface; pipets and burets are usually conditioned.

Constant Composition or Definite Proportions, Law of A given pure compound always contains the same elements in exactly the same proportions by mass.

covalent bond A chemical bond formed by the sharing of electrons between atoms.

Dalton's Law of Partial Pressures Each gas in a mixture of gases exerts a partial pressure equal to the pressure it would exert if it were the only gas present in the same volume; the total pressure of the mixture is then the sum of the partial pressures of all the gases present.

decanting The process of pouring off a supernate into another container.

decomposition reaction Reaction in which two or more elements or compounds are produced from a single substance, usually after the addition of heat.

degree Celsius (°C) A unit of temperature on the Celsius scale.

density Mass of a substance occupying a unit volume.

$$\text{density} = \frac{\text{mass}}{\text{volume}}$$

digestion The process of heating a precipitate in solution in order to develop larger particles free of impurities.

dilute solution   A solution containing a small amount of dissolved solute.

dissociation   The separation of an ionic solute into positive and negative ions by the action of the solvent.

double bond   A bond containing four electrons usually shown as two dashes between two atoms.

double replacement or metathesis reaction   A reaction in which the cations in two different compounds exchange anions.

electromotive series   A sequence of metals arranged in a list according to their ability to undergo reaction. A metal higher in the series will displace another metal from its aqueous solution.

electron-dot formula   A representation of a compound where the electrons in each bond are depicted using dots; also termed Lewis structures.

electronegativity   A measure of the ability of an element to attract electrons.

element   A pure substance that cannot be broken down further by chemical reaction.

empirical formula   The formula of a compound that contains the smallest integral ratio of atoms present in a molecule or formula unit of a compound.

endpoint   The point in a titration when an indicator change is observed.

equivalence point   The point in a titration when the amount of one reactant is equal to the amount of the other reactant according to the balanced chemical equation.

experimental conditions   Conditions of temperature and pressure at which a gas is measured, usually not STP.

experimentation   The collection of data by observation of chemical changes under controlled conditions.

filtrate   The solution that passes through a filtering medium such as filter paper.

filtration   The process of separating supernate from precipitate.

fire polishing   Eliminating rough edges on glass by placing the glass in a hot flame until the glass melts and forms rounded edges.

firing to red heat   Heating a crucible or other porcelain labware until it glows red.

flame cone   The darker blue, conical-shaped inner portion of a flame.

flame envelope   The lighter blue or yellow outer portion of a flame.

flame test   A test for an element's characteristic color when a sample is placed in a hot flame.

formula mass   The mass of a single formula unit, expressed in amu.

formula unit   Generally the smallest combination of charged particles (ions) in which the opposite charges present balance each other so that the overall compound has a net charge of zero, such as NaCl.

freezing point (melting point)   The temperature at which the liquid and solid forms are in dynamic equilibrium with each other. At dynamic equilibrium, the rate of melting is equal to the rate of freezing.

functional group   The atom or group of atoms that defines the structure and properties of a particular class of organic compounds. The site in a molecule where a chemical reaction may occur.

gas   A physical state of matter characterized by large distances between individual particles in random motion. Neither the shape nor the volume is fixed.

gravimetric analysis   A sample analysis in which a weighing operation is used to determine composition.

group   A vertical column in the Periodic Table; also termed *family*.

halide   A compound containing a Group VII A (17) element.

halogens The Group VIIA elements.

heating to constant weight The repeated process of heating, cooling, and weighing until mass readings are constant, or agree closely.

heterogeneous matter Matter not uniform in composition and properties, and consisting of two or more physically distinct portions or phases unevenly distributed.

homogeneous matter Matter uniform in composition and properties throughout.

hydrate salt Crystalline substance that contains chemically bound water in definite proportions.

hydrocarbon derivatives Classes of organic compounds derived from hydrocarbons; examples include alcohols, ethers, aldehydes, ketones, esters.

hydrocarbons Organic compounds consisting of only the elements carbon and hydrogen.

hypothesis A tentative proposal to correlate and explain the experimental data.

immiscible Refers to liquids that are not soluble in one another.

indicator A compound that changes color in an acid or base; example, phenolphthalein.

ionic or electrovalent bond A chemical bond formed by the transfer of one or more electrons from one atom to another.

ionization A process referring to the formation of ions from electrically neutral atoms or molecules.

isomers Compounds that have the same molecular formula but different structural formulas.

IUPAC nomenclature The International Union of Pure and Applied Chemistry (IUPAC) set of systematic rules for naming compounds.

lanthanide series The elements 58 to 71; also referred to as the rare earth elements.

liquid A physical state of matter having no definite shape but fixed volume.

liter (L) A unit of volume corresponding to exactly 1000 cubic centimeters.

luminous flame A cool flame; a flame with a high gas and low oxygen content; also termed a reducing flame.

mass The quantity of matter in a particular body; independent of gravitational attraction; measured by a balance.

mass percent composition An expression for the ratio of the mass of a substance compared to the total mass of the sample, all times 100.

$$\text{mass \%} = \frac{\text{mass substance}}{\text{mass sample}} \times 100$$

melting point (freezing point) The temperature at which the liquid and solid forms are in dynamic equilibrium with each other. At dynamic equilibrium, the rate of melting is equal to the rate of freezing.

meniscus The lens-shaped appearance of the surface of a liquid inside a piece of narrow glassware, such as a pipet or buret.

metal An element possessing a certain set of physical and chemical properties that includes a shiny metallic luster, heat and electrical conductivity, and a tendency to form positive ions.

metric ruler An instrument for measuring length in metric units.

metric system A system of measurement using the standard units of length, mass, and volume: meter, gram, liter.

miscibility Solubility of one liquid in another.

miscible Refers to two liquids that dissolve in one another.

mixture Matter composed of two or more substances, each of which retains its identity and specific properties.

molarity (M) The concentration of solute in a solution expressed as the number of moles of solute per liter of solution.

$$M = \text{molarity} = \frac{\text{moles of solute}}{\text{L of solution}}$$

molar volume of a gas The volume occupied by one mole of any gas; 22.4 L at 0°C and 76$\overline{0}$ torr.

mole The amount of a substance containing the same number of particles, such as atoms, formula units, molecules, or ions, as there are atoms in exactly 12 g of carbon-12. One mole of particles consists of 6.02×10^{23} particles, such as atoms, formula units, molecules, or ions, and this number of particles has a mass equal to the atomic, molecular, or formula mass of the particles expressed in grams.

molecular equation A chemical equation for a reaction in which all species are written as molecules.

molecular formula A formula composed of an appropriate number of symbols of elements representing one molecule of the given compound. Also defined as the true formula and containing the actual number of atoms of each element in one molecule of the compound.

molecular mass The sum of the atomic masses for each atom in the formula of a compound.

net ionic equation A chemical equation containing only those species undergoing a change in the chemical reaction.

neutralization reaction A reaction between an acid and base to form a salt and water.

noble gases The relatively unreactive Group 0 elements; sometimes termed *inert gases*.

nonelectrolyte A substance whose aqueous solutions do not conduct an electric current nor produce a glow in a standard bulb.

nonluminous flame A hot flame; a flame with a high oxygen content; also termed *oxidizing flame*.

nonmetal An element not having metallic properties. Nonmetals ordinarily gain electrons to form negative ions.

nonpolar compound A compound composed of molecules that bear an overall symmetry of positive and negative charge.

octet rule *See* rule of eight.

ordinate axis The vertical axis (y-axis) on a graph.

organic chemistry A study mostly of compounds containing the element carbon.

origin The point of intersection of the abscissa and ordinate axes on a graph.

oxidation A chemical change in which a substance loses electrons or there is an increase in oxidation number.

oxidation number A positive or negative whole number used to describe the combining capacity of an element in a compound.

oxidizing agent The substance reduced in a redox reaction.

parallax An error resulting from observing the meniscus level at an angle other than 90° to a graduated cylinder or other piece of calibrated glassware.

percent by mass concentration The concentration of a solute in a solution expressed as parts by mass of solute per 100 parts by mass of solution.

$$\% \text{ by mass} = \frac{\text{mass of solute}}{\text{mass of solution}} \times 100$$

percent composition *See* mass percent composition.

percent yield The ratio of the actual experimental yield compared to the theoretical yield, times 100.

period A horizontal row in the Periodic Table; also termed *series*.

Periodic Law The physical and chemical properties of the elements are periodic functions of their atomic numbers.

Periodic Table or Periodic Chart The systematic arrangement of the elements that allows elements with similar properties to be grouped together.

physical changes Changes in substances that can be observed without a change in the chemical composition of the substance taking place.

physical properties Properties of substances that can be observed without changing the chemical composition of the substance.

physical state The solid, liquid, or gaseous state of matter.

polar compound A compound whose molecules exist as dipoles; that is, the molecule has regions of partial positive and negative charges.

polyatomic ions Ions consisting of two or more atoms with a net negative or positive charge on the ion.

precipitate (ppt.) An insoluble substance formed from a reaction in solution; exists as solid particles that usually settle from solution.

precision Refers to the spread or range in a set of experimental results. A measure of the reproducibility of a set of results.

qualitative analysis An analysis designed to identify the chemical composition of a substance or solution.

reagent A general term used to describe any chemical substance or solution.

redox reaction A chemical reaction involving the reduction of one species and the oxidation of another.

reducing agent The substance oxidized in a redox reaction.

reduction A chemical change in which a substance gains electrons or there is a decrease in oxidation number.

replacement reaction A reaction in which a more active element displaces a less active element from a compound.

rule of eight In writing electron-dot formulas, each atom in the compound or ion is surrounded by eight electrons. An octet of electrons is known to be energetically stable as in the example of the noble gases. (Hydrogen is an exception to the rule of eight; it shares only two electrons.)

STP *See* standard temperature and pressure.

salt A compound formed when one or more of the hydrogen ions of an acid is replaced by a cation (metal or positive polyatomic ion), or when one or more of the hydroxide ions of a base is replaced by an anion (nonmetal or negative polyatomic ion).

saturated hydrocarbons Hydrocarbons containing only single bonds between carbon atoms; represented by the alkanes.

saturated solution A solution that is in dynamic equilibrium with undissolved solute; that is, the rate of dissolution of undissolved solute is equal to the rate of crystallization of dissolved solute.

$$\text{undissolved solute} \underset{\text{rate of crystallization}}{\overset{\text{rate of dissolution}}{\rightleftharpoons}} \text{dissolved solute}$$

science A study of organized or systematized knowledge.

scientific law   A formal, time-tested, explanation of the correlation of experimental data.

scientific method   The systematic and careful application of experimentation followed by hypothesizing and theorizing.

seed crystal   A small crystal whose chemical composition is similar to the solute. Used to crystallize supersaturated solutions.

semimicro analysis   An analysis using only a few drops of solution to observe reaction.

single bond   A bond composed of two electrons and shown as a single dash.

solid   A physical state of matter having a definite shape and volume.

solubility   The amount of solute that a given quantity of solvent (usually 100 grams) can dissolve at a specified temperature.

solute   The component of a solution that is in lesser quantity.

solution   Homogeneous matter composed of two or more pure substances whose composition can be varied within certain limits.

solvent   The component of a solution that is in greater quantity.

specific gravity   Density of a substance divided by the density of water at 4°C. Specific gravity is a unitless quantity.

specific heat   The number of calories required to raise the temperature of 1.00 g of a substance 1.00°C. The specific heat of water is 1.00 cal/g $\times$ °C.

spectator ions   Ions appearing in the total ionic equation but not participating in the reaction.

standard conditions   *See* standard temperature and pressure.

standardization   The process of establishing the concentration of a solution, usually with a certainty of three or more significant digits.

standard temperature and pressure (STP)   A temperature of 0°C and pressure of $76\overline{0}$ torr arbitrarily defined as standard conditions.

stoichiometry   The quantitative relationship of masses and volumes pertaining to the species in a balanced chemical equation.

strong acid   An acid that is highly ionized in aqueous solution.

strong base   A base that is highly ionized in aqueous solution.

strong electrolyte   A substance whose aqueous solutions conduct an electric current to produce a bright glow in a standard light bulb.

structural formula   A formula showing the arrangement of atoms within a molecule and using a dash for each pair of electrons shared between atoms.

structural isomers   Compounds having the same molecular formulas but different structural formulas.

sublimation   The direct conversion of a solid to the vapor without passing through the liquid state.

substance   Homogeneous matter characterized by definite and constant composition and definite and constant properties under a given set of conditions.

supercooling   The phenomenon demonstrated when a substance is cooled to a temperature below the freezing point while remaining in the liquid state.

supernate   The clear liquid that separates from an insoluble substance in solution.

supersaturated solution   A solution in which the concentration of solute is greater than that possible in a saturated (equilibrium) solution under the same conditions. This solution is unstable and will revert to a saturated solution if a "seed" crystal of solute is added; the excess solute crystallizes out of solution.

theoretical yield   The amount of product theoretically obtained from a reaction after applying a stoichiometric calculation.

theory   A formal proposal that explains the correlation of experimental data.

titration   The process of adding a measured volume of solution through a buret in order to determine the concentration or amount of substance in a solution.

torr   A unit of gas pressure equal to one millimeter of mercury.

total ionic equation   A chemical equation in which all ionized species in solution are written as ions.

transition elements   All of the elements in the B groups and Group VIII of the Periodic Table; the lanthanide and actinide series are referred to as the inner transition elements.

triple bond   A bond containing six electrons shown as three dashes between two atoms.

uncertainty   The degree of inexactness in a measurement due to the limitations of the measuring device. An inherent error in any measurement.

unsaturated hydrocarbons   Hydrocarbons that contain one or more double or triple bonds between carbon atoms; examples include the alkenes and alkynes.

unsaturated solution   A solution in which the concentration of solute is less than that of the saturated (equilibrium) solution under the same conditions.

unshared pair of electrons   A pair of valence electrons (shown in the electron-dot formula) that are not bonded to another atom.

valence electrons   The electrons in the outer energy level; the electrons available for chemical bonding.

vapor pressure   The pressure exerted by the molecules in the vapor (at constant temperature) in dynamic equilibrium with the liquid in a closed system. Dynamic equilibrium is established when the rate of molecules leaving the surface of the liquid (evaporation) is equal to the rate of molecules reentering the liquid (condensation).

volume by displacement   A method for determining volume by the amount of liquid displaced.

volumetric analysis   An analysis of a sample in which a volume measurement is used to determine composition.

water of crystallization   The number of water molecules that crystallize from solution to form a hydrate salt; also termed *water of hydration*.

weak acid   An acid that is weakly ionized in aqueous solution.

weak base   A base that is weakly ionized in aqueous solution.

weak electrolyte   A substance whose aqueous solutions conduct an electric current to produce a dull glow in a standard light bulb.

weighing by difference   A procedure for indirectly obtaining the mass of a sample by weighing a container and then weighing the container with the sample. Samples are not usually weighed directly, a process that would contaminate the balance.

weight   The gravitational force of attraction between a body's mass and the mass of the planet or satellite on which it is weighed; weight is measured by a scale, mass by a balance.

II

Answers to Prelaboratory Assignments

EXPERIMENT 1—SCIENTIFIC OBSERVATIONS

1. See the Glossary, Appendix I.
2. See the diagrams of Common Laboratory Equipment, pages 4–5.
3. Directions for transferring chemicals are given in Appendices IX and X.
4. All chemicals have the potential to be dangerous, as you will see in the experiment.
5. Flush immediately with water and notify the instructor of any irritation.

EXPERIMENT 2—METRIC SYSTEM MEASUREMENTS

1. See the Glossary, Appendix I.
2. See the diagrams of Common Laboratory Equipment, pages 4–5.
3. (a) mass; (b) length; (c) volume; (d) temperature; (e) time
4. (a) ±0.1 g; (b) ±0.01 g; (c) ±0.001 g; (d) ±0.1 cm; (e) ±0.05 cm; (f) ±0.5 mL; (g) ±0.5°C; (h) ±1 s
5. (a) 8.8 cm, 19.1 cm, 7.15 cm, 15.00 cm; (b) 14.0 mL, 84.0 mL; (c) 30.0°C, 2.5°C
6. $11.33125 \approx 11.3 \ cm^3$
7. • A balance is a sensitive instrument and should be handled with care.
   • Be careful to avoid burns from the Bunsen burner, the ring stand, and the boiling waterbath.
   • Thermometers are easily broken and can cause cuts. Always use a thermometer with care. Report a broken thermometer to the instructor and cleanup the mercury immediately.

EXPERIMENT 3—DENSITY OF LIQUIDS AND SOLIDS

1. See the Glossary, Appendix I.
2. (a) 54.0 mL, 82.5 mL; (b) 5.40 cm, 4.15 cm
3. 0.790 g/mL
4. 2.4 g/mL
5. $3.25 \ g/cm^3$
6. $0.00106 \ cm$ or $1.06 \times 10^{-3} \ cm$
7. • Be careful when using the pipet. See Appendix VIII
   • The instructor may wish to point out the disposal of methylene chloride and hexane in a special organic chemical waste container.

EXPERIMENT 4—SPECIFIC HEAT OF A METAL

1. See the Glossary, Appendix I.

2. 100.0 g

3. $1.00 \text{ cal/g} \times {}^{\circ}\text{C}$

4. Two, because the temperature difference has two digits; e.g., 3.5°C.

5. Heat gain water = 800 cal = heat loss metal
   Specific heat Al = $0.202 \text{ cal/g} \times {}^{\circ}\text{C}$

6. • Heat loss from the calorimeter to the surroundings (that is, from the styrofoam cup to the air).
   • Reading the thermometer to 0.1°C.

7. • Handle the laboratory burner and the elevated beaker of boiling water with caution.
   • Carefully transfer the hot metal into the calorimeter water to avoid splashing.

EXPERIMENT 5—PHYSICAL AND CHEMICAL PROPERTIES

1. See the Glossary, Appendix I.

2. See the Discussion section of this experiment.

3. See the Discussion section of this experiment.

4. 2 mL, 5mL

5. To prevent the liquid from "bumping" and spattering from the test tube.

6. All of the following suggest a chemical change: liberating or consuming energy; changing color or texture; forming an insoluble substance in solution; producing a gas.

7. • While determining a boiling point, be careful that the liquid vapors do not come near an open flame.
   • When heating iodine, select a beaker that fits the evaporating dish well.
   • Anytime a chemical is heated in a test tube, the open end should be pointed in a safe direction.
   • Thermometers are easily broken and can cause cuts. Always use a thermometer with great care.

EXPERIMENT 6—CHANGE OF STATE

1. See the Glossary, Appendix I.

2. Heating tap water will leave carbonate deposits in the beaker.

3. The plateau on the cooling curve is extended to the ordinate axis to determine the freezing point of para-dichlorobenzene.

4. Placing the test tube into hot water melts the compound and allows the thermometer to be removed.

5. In taking a melting point it is quite time consuming to heat a substance 1°C per minute. If the approximate melting point is determined rapidly, an accurate value can be found rather quickly in a second trial.

6. $68.0-69.5^{\circ}\text{C}$

7. • The temperature of the waterbath is higher than the Mp of the compound.
   • If the capillary is not sealed completely, water will move into the tube giving the appearance of melting the compound.

8. • Paradichlorobenzene is somewhat flammable and therefore the test tube should not be heated directly with a flame.
   • A thermometer is a delicate instrument and should be handled carefully.

EXPERIMENT 7—PERIODIC CLASSIFICATION OF THE ELEMENTS

1. See the Glossary, Appendix I.

2. Li, Na, and K; Ca, Sr, and Ba; Cl, Br, and I.

3. The wire must be clean. Sodium contamination is everpresent and must be distinguished from a positive test for a sodium compound.

4. The less dense hexane is immiscible with water and forms the upper layer. It is this upper layer that confirms the halide test.

5. • Wear eye protection when flame testing to avoid spattering.
   • Handle acids carefully and avoid breathing the vapors of concentrated hydrochloric acid.
   • Halide test solutions should be emptied into a designated organic waste container.

EXPERIMENT 8—STRUCTURE OF COMPOUNDS

1. See the Glossary, Appendix I.

2. The number of valence electrons correspond to the Group number; thus: 1, 4, 5, 6, 7.

3. (a) hydrogen atom;   (b) carbon atom;   (c) oxygen atom;   (d) single bond (two e^-);
   (e) double bond (four e^-);   (f) triple bond (six e^-)

4. (a) I — Br   : I : Br :

   (b) H   H
        |
     H — C — Cl   H : C : Cl :
        |
        H   H

   (c) O   : O
        ||
    Cl — C — Cl   : Cl : C : Cl :

5. (a) $7 + 7 = 14e^-$ (b) $3 + 4 + 7 = 14e^-$ (c) $14 + 4 + 6 = 24e^-$

EXPERIMENT 9—CATION ANALYSIS

1. See the Glossary, Appendix I.

2. Tap water contains many ions, some of which can interfere with the results of the analysis.

3. Add a drop of precipitating reagent to the clear supernate. If it becomes cloudy, precipitation is incomplete.

4. Place a stirring rod in the test solution; touch it to red litmus paper. If the paper turns blue, the solution is basic.

5. • Goggles should always be worn when performing a flame test.
   • Contact with dilute HCl and NaOH should be avoided. Any irritation should be immediately flushed with water.
   • The centrifuge must be balanced before operating.

6. Ba^{2+}, Ca^{2+}, and Mg^{2+} are all confirmed.

EXPERIMENT 10—ANION ANALYSIS

1. See the Glossary, Appendix I.

2. Tap water contains many ions, some of which can interfere with the results of the analysis.

3. Add several drops of distilled water, mix with a stirring rod, centrifuge, and discard the decantate.

4. Place a stirring rod in the test solution; touch it to blue litmus paper. If the paper turns red, the solution is acidic.

5. • Contact with dilute HNO_3 and $NH_3 \cdot H_2O$ should be avoided. Any irritation should be immediately flushed with water. $AgNO_3$ will stain your skin.
   • The centrifuge must be balanced before operating.

6. I^{1-}, Cl^{1-}, and $SO_4{}^{2-}$ are all confirmed.

EXPERIMENT 11—AVOGADRO'S NUMBER

1. See the Glossary, Appendix I.

2. —COOH.

3. 177 cm^2; 21 Å^2; 8.4 × 10^{16} molecules/monolayer.

4. The bead or clear lens disappears.

5. When the clear lens persists for 30 seconds.

6. • Error is introduced by delivering drops of solution that are too large or vary in size.
   • It is most important that the surface of the watchglass is clean. Touching the surface can leave a trace of oil that dissolves the stearic acid and leads to meaningless results.

7. • The point of the dropper pipet is quite sharp.
   • Avoid breathing the organic vapor.
   • The hexane solution is quite flammable. Unused portions of the hexane solution should be emptied into a designated organic waste container.

EXPERIMENT 12—PERCENTAGE OF WATER IN A HYDRATE

1. See the Glossary, Appendix I.

2. (d) 96.818 g. The sample is approximately 1.4 g; within the suggested range of 1.2-1.8 g. The mass is recorded to the nearest milligram, the same uncertainty as the beaker plus watchglass.

3. All moisture is absent and the hydrate salt no longer appears crystalline.

4. A warm beaker will warm the air around the balance creating a buoyancy effect. The masses will weigh light.

5. 0.169 g/1.192 g × 100 = 14.2%. The experimental result agrees well with the theoretical value, 14.7%.

6. • Remember that hot glass and cool glass look alike. Allow 10 minutes for the beaker to cool.
   • Take care to avoid breaking the watchglass. It may be necessary to remove the watchglass and crucible tongs at the end of the heating to remove the last traces of water vapor.

EXPERIMENT 13—EMPIRICAL FORMULA

1. See the Glossary, Appendix I.

2. The crucible tongs eliminate weighing errors owing to fingerprints. The crucible should be cool before weighing so a hot crucible is not a problem.

3. The suggested periods for heating and cooling are intended as general guidelines. More important is consistency in the time intervals.

4. Heating magnesium in air produces both magnesium nitride and magnesium oxide. Water reacts with magnesium nitride producing ammonia gas and magnesium hydroxide. Reheating the crucible converts magnesium hydroxide to magnesium oxide.

5. If the magnesium has not completely reacted, small sparks will be observed when the crucible cover is lifted.

6. The reaction of the copper and sulfur is complete when no sulfur is left in the crucible. If in doubt, you can bring the crucible to constant weight.

7. • A hot crucible creates a buoyancy effect on the balance and mass readings are low.
   • The magnesium must not smoke excessively nor spark when the crucible cover is lifted.
   • The sulfur has a tendency to "creep" out of the crucible when heated. Any excess sulfur on the crucible must be heated and driven off as a gas.

8.
- Before firing to red heat, set the crucible on the lab bench and strike sharply with a pencil. A crucible with a hairline crack does not give a clear ring.
- A crucible at red heat has a temperature in excess of $1100^{\circ}C$. Below this temperature a crucible may not glow red; however, it can cause a painful burn. The crucible cover must be handled with tongs.
- The magnesium ignition is strongly exothermic and frequently cracks crucibles. A few porcelain chips in the bottom of the crucible under the magnesium may minimize this problem.
- Heating copper and sulfur together also produces toxic sulfur dioxide gas. This gas must be vented properly; avoid breathing the gas.
- Wearing safety goggles is a necessity when lifting the crucible cover to check the progress of reaction.

EXPERIMENT 14—CHEMICAL REACTIONS

1. See the Glossary, Appendix I.
2. Heat catalyst, no reaction, solid or precipitate, liquid, gas, aqueous solution.
3.
   - A gas is produced.
   - A precipitate is formed.
   - A color change is observed.
   - A temperature change is noted.
4. 2 mL
5. Colorless; pink
6.
   - Ignited magnesium ribbon produces intense heat.
   - Heating a mixture of powdered zinc and sulfur gives a highly exothermic reaction.
   - Avoid the fumes from hydrochloric acid.
7. 78.8 g/mole

EXPERIMENT 15—ANALYSIS BY DECOMPOSITION

1. See the Glossary, Appendix I.
2. Manganese dioxide catalyzes a smooth and safe decomposition of potassium chlorate.
3. The mass of oxygen is found from the difference in the mass of test tube and contents before and after heating.
4. When all of the potassium chlorate is decomposed, no more oxygen is produced and the water level in the beaker remains constant. After the burner is shut off, the water level will actually decrease in the beaker as the oxygen gas is allowed to cool.
5.
   - Insufficient heating of the potassium chlorate mixture.
   - Weighing the test tube while warm; this creates a buoyancy effect and consequently light weighings.
6. Yes, insufficient heating can give a percent yield greater than 100%.
7.
   - The rule to wear safety goggles should be strictly obeyed throughout the experiment.
   - The potassium chlorate mixtures must not contact the rubber stopper in the test tube as a vigorous reaction could occur.
   - Potassium chlorate should not be heated without manganese dioxide to moderate the reaction.
   - Potassium chlorate is a white salt; manganese dioxide a black powder; mixtures are typically gray.
   - It is hazardous to heat a closed gaseous system. Never pinch off the tubing leading from the test tube or flask.

EXPERIMENT 16—ANALYSIS BY PRECIPITATION

1. See the Glossary, Appendix I.
2. Particles of precipitate will be transferred prematurely into the filter paper and thus slow the filtration rate.
3. To thoroughly remove the precipitate from the inside of the beaker.
4. On occasion, particles of precipitate appear in the filtrate. This is due to either a small hole in the filter paper or overfilling the filter paper above the torn corner. At the discretion of the instructor, you may have to recycle the filtrate through a second weighed circle of filter paper. In that event, add together the masses of the two separate precipitates.

5.   • Incomplete precipitation (low results).
     • Coprecipitation of impurities (high results).
     • Precipitate in filtrate (low results).
     • Weighing filter paper with precipitate before it is completely dry (high results).

6.   Yes, co-precipitation or wet filter paper can give a percent yield above 100%.

7.   Transferring the precipitate from the beaker to funnel should be performed carefully to avoid breaking glassware.

EXPERIMENT 17—MOLAR VOLUME OF A GAS

1.   See the Glossary, Appendix I.

2.   A gas rises in a container filled with water. The volume of gas collected is equal to the volume of water displaced.

3.   If the magnesium has a mass greater than 0.09 g, it may produce a volume of hydrogen in excess of the capacity of the 100 mL graduated cylinder.

4.   Gas bubbles are no longer released.

5. $23°C$.

6.   The atmospheric pressure is 756 torr and the vapor pressure of water at $23°C$ is 21.1 torr. Therefore, the partial pressure of hydrogen is $756 - 21.1 = 735$ torr.

7.   The sources of error include air bubbles in the graduated cylinder before reaction, incomplete reaction of the metal, and weighing errors. The most common error is misreading the meniscus in the graduated cylinder after reaction.

8.   • Pour the dilute hydrochloric acid cautiously and avoid breathing the fumes.
     • Carefully handle all glassware to avoid breakage.

EXPERIMENT 18—MOLECULAR MASS OF A GAS

1.   See the Glossary, Appendix I.

2.   Place the rubber tubing in the bottom of the 250 mL flask, flush for five seconds, stopper immediately, and weigh. Repeat this procedure, reweigh, and check the two mass readings for agreement.

3.   A gas having a density less than air (1.3 g/mL) will rise. Therefore, the gas should be collected in an inverted flask.

4.   75.8 cm Hg

5.   The approximate mass of water is $387 - 122 = 265$ g. Since the density of water is 1.00 g/mL, the volume of water is 265 mL. Thus, the volume of the flask is 265 mL and contains 265 mL of air or any other gas.

6.   • Obtaining a representative sample of gas in the flask demands attention to detail. For example, flushing air from the flask must be performed completely and stoppered immediately.
     • The masses of the gas samples in this experiment are on the order of a few tenths of a gram, so the weighing operation is critical. Any moisture in the flask will cause serious errors for the mass readings.

7.   Some of the gases used in this experiment are flammable or foul-smelling. Natural gas is flammable and has an unpleasant odor. Do not have an open flame in the laboratory. Foul-smelling gases may be used under a fume hood.

EXPERIMENT 19—SOLUTIONS

1.   See the Glossary, Appendix I.

2.   Polar solvents dissolve ionic (polar) compounds and nonpolar solvents dissolve covalent (nonpolar) compounds.

3.   Assuming the crystal has color, the solvent will become lightly colored.

4.   Heating tap water leaves carbonate deposits on glassware.

5.   To provide a homogeneous solution; otherwise, the concentration is greater at the bottom of the test tube. Jarring the test tube, after cooling, may cause the solution to crystallize prematurely.

6.   (a) 4.50%;   (b) 0.793 M.

7.   • Pipetting is the major source of error. A 10.0 mL sample of solution should have a mass greater than 10.0 g. If it does not, check your data and repipet if necessary.
     • After evaporating the solution to a dry residue, the dish must be heated to remove all traces of moisture.

8.   • To dispose of the organic chemicals used in Procedure A, the Instructor may wish to provide a waste container. Avoid contact and vapors of all organic chemicals.
     • Heating a beaker of water on the ring stand and pipetting should be performed carefully.
     • When inserting the pipet into the pipet bulb, lubricate the bulb and hold your hands close together.
     • Organic liquids are flammable (methanol, ethanol, acetone, hexane, heptane).

EXPERIMENT 20—NEUTRALIZATION TITRATION

1.   See the Glossary, Appendix I.

2.   (a), (b), (c), (f)

3.   (a) 0.50 mL;   (b) 31.35 mL

4.   Flashes of pink indicator persist longer. One drop, about 0.05 mL, will flip the indicator to permanent pink at the endpoint.

5.   The volume of base will vary depending upon the sample size of KHP.

6.   The volume of base should be consistent (30.15 mL) since the amount of vinegar is constant (10.0 mL).

7.   • Sodium hydroxide is caustic and could be severely harmful if a drop got in your eyes. On your skin, it creates a slippery sensation. In either event, wash immediately with water and notify the instructor.
     • Be especially careful not to overfill the buret by adding too much base into the funnel.
     • When inserting the pipet into the pipet bulb, hold your hands very close together to avoid snapping the pipet.
     • The pipet and buret are expensive pieces of glassware and should be handled with care.

EXPERIMENT 21—IONIC EQUATIONS

1.   See the Glossary, Appendix I.

2.   (g) indicates a gas; (ℓ) a pure liquid; (s) a solid substance; (aq) an aqueous solution.

3.   See Table 21-1.

4.   (a) Bulb glows brightly.
     (b) Bulb has a dull glow.
     (c) Bulb does not glow.

5.   (a) $K^{1+}_{(aq)} + Cl^{1-}_{(aq)}$;   (b) $C_3H_5OH_{(aq)}$;
     (c) $HNO_{2(aq)}$

6.   Any contamination of ions will give a strong conductivity test even for a weak electrolyte or nonelectrolyte.

7.   • Do not touch the exposed wire electrodes; a very serious shock can result.
     • Any chemicals contacting your skin or clothes should be washed immediately.

EXPERIMENT 22—OXIDATION-REDUCTION

1.   See the Glossary, Appendix I.

2.   (a) 0, (b) −1, (c) +1, (d) +7, (e) +5

3.   (a) $Cr_2O_7^{2-} + 6\ Fe^{2+} + 14\ H^{1+} \longrightarrow 2\ Cr^{3+} + 6\ Fe^{3+} + 7\ H_2O$
     (b) $Mn^{2+} + H_2O_2 + 2\ OH^{1-} \longrightarrow MnO_2 + 2\ H_2O$

264   ANSWERS TO PRELABORATORY ASSIGNMENTS

4. Al, Cd, Ni, Ag

5. • Concentrated nitric acid is a very strong acid and should be handled with caution. If contact occurs, wash immediately with water and notify the laboratory instructor.

    • All of the following gases are highly irritating and should be observed carefully with proper ventilation: SO_2, H_2S, NO, NO_2, and NH_3.

    • Always pour the concentrated solution into water in order to dilute the concentration.

EXPERIMENT 23—ORGANIC MODEL EXERCISE

1. See the Glossary, Appendix I.

2. The element represented is carbon, oxygen, hydrogen, chlorine, bromine, iodine, and nitrogen, respectively.

3. Two black balls and a single connector.

4. A black ball and a red ball joined by two connectors.

5. (a) ether;   (b) organic halide;   (c) alkene;   (d) ketone;   (e) carboxylic acid;
    (f) alkyne;   (g) amide;   (h) alcohol;   (i) phenol;   (j) ester;   (k) amine;
    (l) aldehyde

6. (a) isopropyl;   (b) propyl;   (c) methyl;   (d) ethyl

7.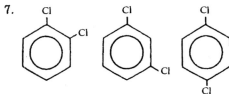

    ortho   meta   para

Using a Laboratory Burner

Although a variety of burners are found in chemistry laboratories, they all employ the same principle. Natural gas is allowed to flow into the barrel of the burner and mix with the air which contains oxygen. The ratio of gas to air can be adjusted, which in turn regulates the temperature of the flame. The more air that is available, the hotter the flame. Two typical burners are shown in Figure III-1.

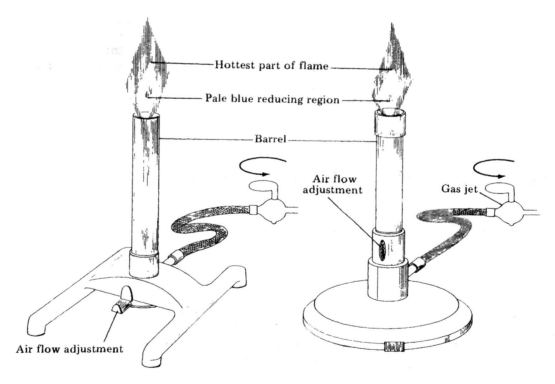

FIGURE III-1　Laboratory burners.

Steps in Operating a Burner

1. Close the air flow adjustment.
2. Open the gas jet.
3. Light the burner at the top of the barrel.
4. To obtain a hotter flame, open the air flow adjustment.
5. To shut off the burner, close the gas jet.

Manipulating Glass Tubing

Cutting Glass Tubing   Figure IV-1 illustrates this process. Measure a 15 cm length of 6 mm (OD) glass tubing. Place the tubing flat on the laboratory bench and firmly press a triangular file against the tubing. Push the file away from you, making a single stroke. If the file is dull, a few more strokes may be necessary to make a good scratch; however, there should be only a single groove and the scratch need not be very long or deep. Grasp the tubing and rotate it so that the scratch is opposite your thumbs. You should be able to break the tubing easily and cleanly by pushing out with your thumbs and pulling back lightly with your fingers. If a small glass spur remains after breaking, use the file to remove it.

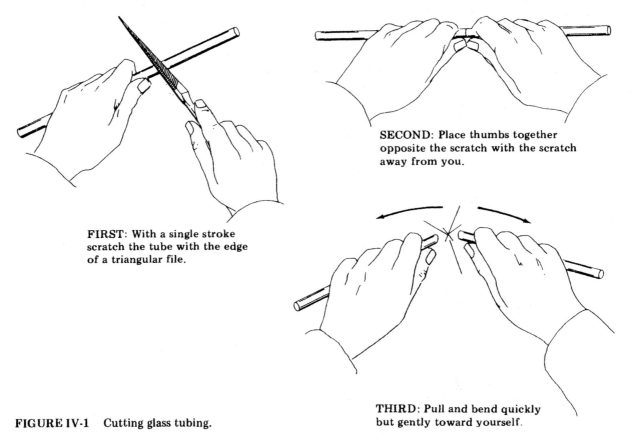

FIRST: With a single stroke scratch the tube with the edge of a triangular file.

SECOND: Place thumbs together opposite the scratch with the scratch away from you.

THIRD: Pull and bend quickly but gently toward yourself.

FIGURE IV-1   Cutting glass tubing.

Fire Polishing   After cutting glass tubing, the resulting sharp edges present a hazard. The technique of smoothing a sharp edge by heating in a flame is termed fire polishing.

Fire polish both ends of the 15 cm piece of tubing (Figure IV-2). First, place the end of the tubing along the inner cone at an angle of 45°. Rotate the tubing in the flame until the edge is smooth. The inside diameter should not become constricted.

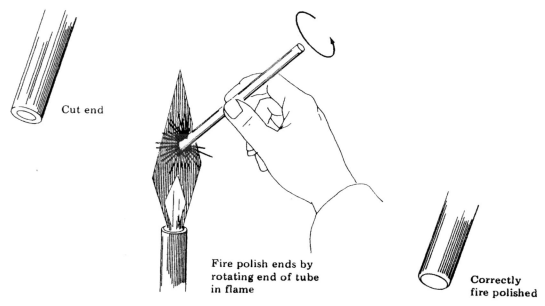

Cut end

Fire polish ends by rotating end of tube in flame

Correctly fire polished

FIGURE IV-2   Fire polishing.

CAUTION: To avoid a burn from hot glass, remember that hot and cool glass appear the same. If you would like the instructor to examine your glassworking, do not offer an end that has just been removed from the flame. Some instructors have reportedly reacted unpleasantly to handling a hot piece of glass tubing.

Glassworking   Shut off the burner and place a wing tip (flame spreader) on the barrel. Heat the 15 cm piece of tubing as shown in Figure IV-3 and prepare a 90° bend.

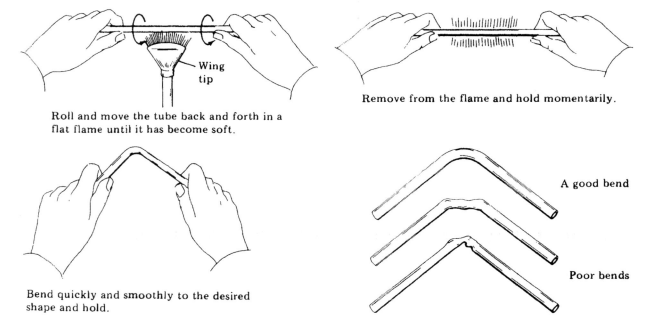

Wing tip

Roll and move the tube back and forth in a flat flame until it has become soft.

Remove from the flame and hold momentarily.

Bend quickly and smoothly to the desired shape and hold.

A good bend

Poor bends

FIGURE IV-3   Bending glass tubing.

Making a Dropper Pipet   Cut another 15 cm section of glass tubing and fire polish both ends. Prepare a dropper pipet as shown in Figure IV-4.

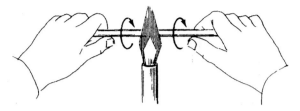

Roll the tube in a flame until it softens.

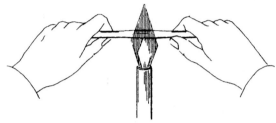

Allow the tube to shorten and thicken slightly as it heats.

Remove from the flame and pull until the capillary is as small as desired.

Cut to length and fire polish the ends.

FIGURE IV-4   Preparing a dropper pipet.

Using a Platform Balance

The platform balance is used to obtain the mass of objects where a precision of one decigram (0.1 g) is acceptable (Figure V-1). By adjusting the riders on the beams of the balance, the mass on the pan can be counterbalanced. When the pointer at the end of the beam is aligned with the zero point, the mass of the sample can be determined by adding up the readings of the riders on the balance beam.

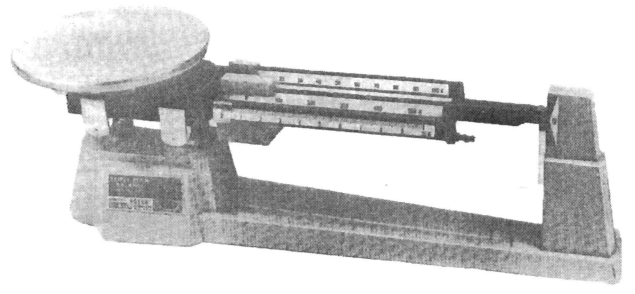

FIGURE V-1   A platform balance with decigram precision (0.1g). (Photo courtesy of Ohaus Scale Corporation.)

Using a Centigram Balance

A centigram balance is satisfactory for mass measurements where a precision of one centigram (0.01 g) is acceptable (Figure VI-1). By adjusting the riders on the balance beams, the mass on the pan can be determined. When the pointer at the end of the beam points to zero, the mass of the sample is equal to the sum of the masses indicated by the riders on the balance beams.

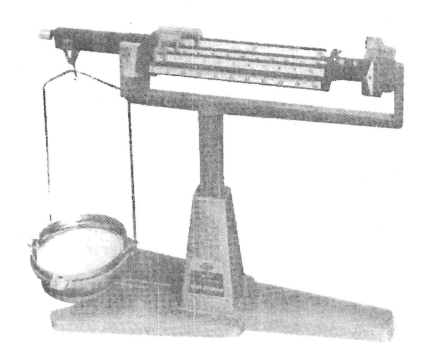

FIGURE VI-1   A centigram balance with a precision of 0.01g. (Photo courtesy of Ohaus Scale Corporation.)

Using an Electronic Balance

An electronic balance is distinctly more sophisticated than the platform or centigram balance. This balance is an *expensive* electronic instrument that should be carefully used only after instruction in its operation. There are a variety of top-loading models available and most have milligram (0.001 g) precision (Figure VII-1).

Although it is not readily obvious, the principle of operation is similar to the less sophisticated beam balances. By rotating the weight adjustment knob, riders are substituted on and off the counterbalance beam which is inside the balance case and not visible. Some of the top-loading balances have an additional adjustment knob for a more precise reading of the milligram digit.

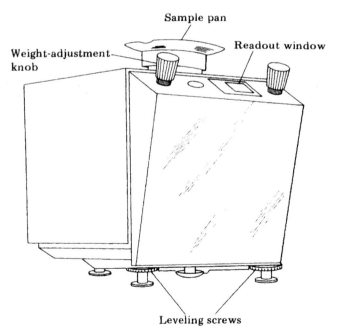

FIGURE VII-1   A top-loading electronic balance with milligram precision (0.001 g).

Using a Volumetric Pipet

In using a volumetric pipet, the following steps are useful to remember.

1. Lubricate the bulb before inserting the pipet. Hold your hands close together so as to avoid snapping the pipet, thereby avoiding a possible accident.
2. If the pipet is not dry, condition the pipet with a portion of the solution to be transferred. Dispose of the conditioning solution in the sink, unless otherwise directed.
3. Use the pipet bulb *(not your mouth)* to draw up the solution above the calibration line. Slip off the bulb and place your finger on the end of the pipet, thus preventing the solution from draining. Carefully allow the bottom of the meniscus to drop to the calibration line by moving your finger slightly.
4. Touch off the last drop of solution at the tip of the pipet and transfer the tip of the pipet into the receiving flask or beaker. Allow the solution to drain free into the receiving flask. Touch off the tip of the pipet. Do not blow out the last drop of solution, as the pipet is calibrated to deliver the specified amount (Figure VIII-1).

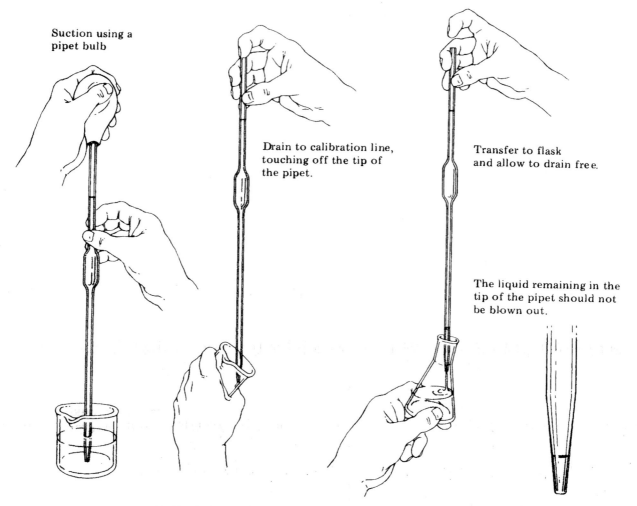

Suction using a
pipet bulb

Drain to calibration line,
touching off the tip of
the pipet.

Transfer to flask
and allow to drain free.

The liquid remaining in the
tip of the pipet should not
be blown out.

FIGURE VIII-1 Transferring a quantity of solution using a
volumetric pipet and bulb.

IX

Transferring a Solid from a Reagent Bottle

Please observe the following general procedures.

1. Always obtain samples of crystals or powders at the shelf where the bottle is stored. Do not take the bottle to *your* laboratory station.

   NOTE: If an experiment calls for a *mass* of reagent, it is assumed a solid is being designated, as opposed to a liquid. A gram of substance is (very roughly) about a pea-sized portion.

2. Remove the glass stopper and scoop out a small portion of substance (Figure IX-1). Tap the scoopula with a pencil or your finger until the desired amount falls onto a watchglass, beaker, or other receiving vessel. Unless you are instructed otherwise, return the excess amount of substance on the scoopula to the reagent bottle.

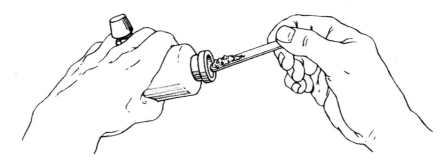

FIGURE IX-1   Scoop out a little of the material with a lab scoopula.

3. Place the same stopper back into the reagent bottle and return the bottle to its proper position on the shelf.
4. For larger amounts where a coarse measurement is tolerable, simply pour the solid into a beaker with a rolling motion from the reagent bottle.

Transferring a Liquid from a Reagent Bottle

Please observe the following general procedures.

1. Always obtain samples of liquids at the shelf where the bottle is stored. Do not take the bottle to *your* laboratory station.

   NOTE: If an experiment calls for a *volume* of reagent, it is assumed a liquid is being designated, rather than a solid. A standard 16 X 150 mm test tube contains about 20 mL. A standard dropper pipet delivers about 20 drops per milliliter.

2. Remove the glass stopper and hold the stopper between two fingers as shown in Figure X-1. Never set a stopper down. Pouring the liquid down the side of a glass rod minimizes splashing.

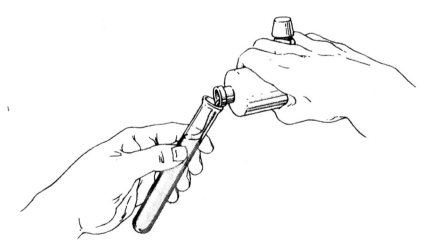

FIGURE X-1   Pouring a liquid directly from a reagent bottle.

3. Narrow glassware such as graduated cylinders and burets do not lend themselves to the use of a stirring rod. In such cases pour directly into the vessel (Figure X-1) or pour through a funnel.

4. Place the stopper back into the reagent bottle and return the bottle to its proper position on the reagent shelf.

Common Cations, Anions, and Polyatomic Ions

Cation	Name of Cation
Al^{3+}	Aluminum
Ba^{2+}	Barium
Bi^{3+}	Bismuth
Cd^{2+}	Cadmium
Ca^{2+}	Calcium
Cu^{1+}	Copper(I) or cuprous
Cu^{2+}	Copper(II) or cupric
Au^{3+}	Gold(III)
H^{1+}	Hydrogen
Fe^{2+}	Iron(II) or ferrous
Fe^{3+}	Iron(III) or ferric
Pb^{2+}	Lead(II) or plumbous
Pb^{4+}	Lead(IV) or plumbic
Li^{1+}	Lithium
Mg^{2+}	Magnesium
Hg_2^{2+}	Mercury(I) or mercurous
Hg^{2+}	Mercury(II) or mercuric
Ni^{2+}	Nickel(II)
K^{1+}	Potassium
Ag^{1+}	Silver
Na^{1+}	Sodium
Sr^{2+}	Strontium
Sn^{2+}	Tin(II) or stannous
Sn^{4+}	Tin(IV) or stannic
Zn^{2+}	Zinc

Anion	Name of Anion
Br^{1-}	Bromide
Cl^{1-}	Chloride
F^{1-}	Fluoride
H^{1-}	Hydride
I^{1-}	Iodide
N^{3-}	Nitride
O^{2-}	Oxide
P^{3-}	Phosphide
S^{2-}	Sulfide

Polyatomic Ion	Name of Polyatomic Ion
$C_2H_3O_2^{1-}$	Acetate
NH_4^{1+}	Ammonium
CO_3^{2-}	Carbonate
ClO_3^{1-}	Chlorate
ClO_2^{1-}	Chlorite
CrO_4^{2-}	Chromate
CN^{1-}	Cyanide
$Cr_2O_7^{2-}$	Dichromate
HCO_3^{1-}	Hydrogen carbonate or bicarbonate
HSO_4^{1-}	Hydrogen sulfate or bisulfate
HSO_3^{1-}	Hydrogen sulfite or bisulfite
OH^{1-}	Hydroxide
ClO^{1-}	Hypochlorite
NO_3^{1-}	Nitrate
NO_2^{1-}	Nitrite
$C_2O_4^{2-}$	Oxalate
ClO_4^{1-}	Perchlorate
MnO_4^{1-}	Permanganate
PO_4^{3-}	Phosphate
SO_4^{2-}	Sulfate
SO_3^{2-}	Sulfite

Generalizations
of the Solubility
# of Solids in Water

1. Nearly all **nitrates** and **acetates** are *soluble.*
2. All **chlorides** are *soluble* except AgCl, Hg_2Cl_2 and $PbCl_2$. ($PbCl_2$ is soluble in hot water.)
3. All **sulfates** are *soluble* except $BaSO_4$, $SrSO_4$, and $PbSO_4$. ($CaSO_4$ and Ag_2SO_4 are only slightly soluble.)
4. Most of the **alkali metal** (Li, Na, K, etc.) salts and **ammonium** salts are *soluble.*
5. All **oxides** and **hydroxides** are *insoluble* except those of the alkali metals, and certain alkaline earth metals (Ca, Sr, Ba, Ra). [$Ca(OH)_2$ is only moderately soluble.]
6. All **sulfides** are *insoluble* except those of the alkali metals, alkaline earth metals, and ammonium sulfide.
7. All **phosphates** and **carbonates** are *insoluble* except those of the alkali metals and ammonium salts.

Electromotive Series

Li
K
Ba
Ca
Na
Mg
Al
Zn
Fe
Cd
Ni
Sn
Pb
(H)
Cu
Hg
Ag
Au

Concentrations of Dilute and Concentrated Acids and Bases

Reagent	Formula	Molar Concentration	Percent by Mass Concentration	Specific Gravity
acetic acid, conc	$HC_2H_3O_2$	17 M	99.5%	1.05
acetic acid, dil		6 M	34%	1.04
hydrochloric acid, conc	HCl	12 M	36%	1.18
hydrochloric acid, dil		6 M	20%	1.10
nitric acid, conc	HNO_3	16 M	72%	1.42
nitric acid, dil		6 M	32%	1.19
sulfuric acid, conc	H_2SO_4	18 M	96%	1.84
sulfuric acid, dil		3 M	25%	1.18
ammonia water, conc	$NH_3 \cdot H_2O$	15 M	58%	0.90
ammonia water, dil		6 M	23%	0.96
sodium hydroxide, dil	NaOH	6 M	20%	1.22

List of Elements with Their Symbols and Atomic Masses

Element	Symbol	Atomic Number	Atomic Mass[a] (amu)
Actinium	Ac	89	(227)
Aluminum	Al	13	26.9815
Americium	Am	95	(243)
Antimony	Sb	51	121.75
Argon	Ar	18	39.948
Arsenic	As	33	74.9216
Astatine	At	85	(210)
Barium	Ba	56	137.33
Berkelium	Bk	97	(247)
Beryllium	Be	4	9.01218
Bismuth	Bi	83	208.9806
Boron	B	5	10.811
Bromine	Br	35	79.904
Cadmium	Cd	48	112.41
Calcium	Ca	20	40.08
Californium	Cf	98	(251)
Carbon	C	6	12.01115
Cerium	Ce	58	140.12
Cesium	Cs	55	132.9055
Chlorine	Cl	17	35.453
Chromium	Cr	24	51.996
Cobalt	Co	27	58.9332
Copper	Cu	29	63.546
Curium	Cm	96	(247)
Dysprosium	Dy	66	162.50
Einsteinium	Es	99	(254)
Erbium	Er	68	167.26
Europium	Eu	63	151.96
Fermium	Fm	100	(257)
Fluorine	F	9	18.998403
Francium	Fr	87	(223)
Gadolinium	Gd	64	157.25
Gallium	Ga	31	69.72
Germanium	Ge	32	72.59
Gold	Au	79	196.9665
Hafnium	Hf	72	178.49
Hahnium[b]	Ha	105	(262)
Helium	He	2	4.00260
Holmium	Ho	67	164.9303
Hydrogen	H	1	1.0080
Indium	In	49	114.82
Iodine	I	53	126.9045
Iridium	Ir	77	192.22
Iron	Fe	26	55.847
Krypton	Kr	36	83.80
Lanthanum	La	57	138.9055
Lawrencium	Lr	103	(257)
Lead	Pb	82	207.2
Lithium	Li	3	6.941
Lutetium	Lu	71	174.967
Magnesium	Mg	12	24.305
Manganese	Mn	25	54.9380
Mendelevium	Md	101	(256)
Mercury	Hg	80	200.59
Molybdenum	Mo	42	95.94
Neodymium	Nd	60	144.24
Neon	Ne	10	20.179
Neptunium	Np	93	237.0482
Nickel	Ni	28	58.70
Niobium	Nb	41	92.9064
Nitrogen	N	7	14.0067
Nobelium	No	102	(255)
Osmium	Os	76	190.2
Oxygen	O	8	15.9994
Palladium	Pd	46	106.4
Phosphorus	P	15	30.9738
Platinum	Pt	78	195.09
Plutonium	Pu	94	(244)
Polonium	Po	84	(209)
Potassium	K	19	39.0983
Praseodymium	Pr	59	140.9077
Promethium	Pm	61	(145)
Protactinium	Pa	91	231.0359
Radium	Ra	88	226.0254
Radon	Rn	86	(222)
Rhenium	Re	75	186.207
Rhodium	Rh	45	102.9055
Rubidium	Rb	37	85.4678
Ruthenium	Ru	44	101.07
Rutherfordium[b]	Rf	104	(261)
Samarium	Sm	62	150.4
Scandium	Sc	21	44.9559
Selenium	Se	34	78.96
Silicon	Si	14	28.0855
Silver	Ag	47	107.868
Sodium	Na	11	22.9898
Strontium	Sr	38	87.62
Sulfur	S	16	32.06
Tantalum	Ta	73	180.9479
Technetium	Tc	43	98.9062
Tellurium	Te	52	127.60
Terbium	Tb	65	158.9254
Thallium	Tl	81	204.37
Thorium	Th	90	232.0381
Thulium	Tm	69	168.9342
Tin	Sn	50	118.69
Titanium	Ti	22	47.90
Tungsten	W	74	183.85
Unnilennium	Une	109	(266)
Unnilhexium	Unh	106	(263)
Unniloctium	Uno	108	(265)
Unnilseptium	Uns	107	(262)
Uranium	U	92	238.029
Vanadium	V	23	50.9415
Xenon	Xe	54	131.30
Ytterbium	Yb	70	173.04
Yttrium	Y	39	88.9059
Zinc	Zn	30	65.37
Zirconium	Zr	40	91.22

[a]Based on the assigned relative atomic mass of ^{12}C = exactly 12 amu; [b]name and symbol not officially approved; parentheses indicate the mass number of the isotope with the longest half-life.

Periodic Table of the Elements

GROUPS

PERIODS	1 IA	2 IIA	3 IIIB	4 IVB	5 VB	6 VIB	7 VIIB	8 VIII	9 VIII	10 VIII	11 IB	12 IIB	13 IIIA	14 IVA	15 VA	16 VIA	17 VIIA	18 VIIIA
1	1.008 H 1																	4.003 He 2
2	6.941 Li 3	9.012 Be 4											10.811 B 5	12.011 C 6	14.007 N 7	15.999 O 8	18.998 F 9	20.179 Ne 10
3	22.990 Na 11	24.305 Mg 12											26.982 Al 13	28.0855 Si 14	30.9738 P 15	32.06 S 16	35.453 Cl 17	39.948 Ar 18
4	39.0983 K 19	40.08 Ca 20	44.956 Sc 21	47.90 Ti 22	50.9415 V 23	51.996 Cr 24	54.938 Mn 25	55.847 Fe 26	58.933 Co 27	58.71 Ni 28	63.546 Cu 29	65.37 Zn 30	69.72 Ga 31	72.59 Ge 32	74.922 As 33	78.96 Se 34	79.904 Br 35	83.80 Kr 36
5	85.468 Rb 37	87.62 Sr 38	88.906 Y 39	91.22 Zr 40	92.9064 Nb 41	95.94 Mo 42	98.906 Tc 43	101.07 Ru 44	102.906 Rh 45	106.4 Pd 46	107.868 Ag 47	112.41 Cd 48	114.82 In 49	118.69 Sn 50	121.75 Sb 51	127.60 Te 52	126.904 I 53	131.30 Xe 54
6	132.906 Cs 55	137.33 Ba 56	138.906 *La 57	178.49 Hf 72	180.948 Ta 73	183.85 W 74	186.2 Re 75	190.2 Os 76	192.22 Ir 77	195.09 Pt 78	196.967 Au 79	200.59 Hg 80	204.37 Tl 81	207.2 Pb 82	208.981 Bi 83	(209) Po 84	(210) At 85	(222) Rn 86
7	(223) Fr 87	226.025 Ra 88	(227) **Ac 89	(261) Rf 104	(262) Ha 105	(263) Unh 106	(262) Uns 107	(265) Uno 108	(266) Une 109									

TRANSITION ELEMENTS

*Lanthanide series

140.12 Ce 58	140.908 Pr 59	144.24 Nd 60	(145) Pm 61	150.4 Sm 62	151.96 Eu 63	157.25 Gd 64	158.925 Tb 65	162.50 Dy 66	164.930 Ho 67	167.26 Er 68	168.934 Tm 69	173.04 Yb 70	174.967 Lu 71

**Actinide series

232.038 Th 90	231.031 Pa 91	238.029 U 92	237.048 Np 93	(244) Pu 94	(243) Am 95	(247) Cm 96	(247) Bk 97	(251) Cf 98	(254) Es 99	(257) Fm 100	(256) Md 101	(255) No 102	(257) Lr 103

Numbers below the symbol of the element indicate the atomic numbers. Atomic masses, above the symbol of the element, are based on the assigned relative atomic mass of ^{12}C = exactly 12; ( ) indicates the mass number of the isotope with the longest half-life.

282

Notes

Notes

Notes

Notes

Notes

Notes

Notes

Notes